Language, Development Aid and Human Rights in Education

Palgrave Studies in Global Citizenship Education and Democracy

Series Editor: **Jason Laker**, San Jose University, USA

This series will engage with the theoretical and practical debates regarding citizenship, human rights education, social inclusion, and individual and group identities as they relate to the role of higher and adult education on an international scale. Books in the series will consider hopeful possibilities for the capacity of higher and adult education to enable citizenship, human rights, democracy and the common good, including emerging research and interesting and effective practices. It will also participate in and stimulate deliberation and debate about the constraints, barriers and sources and forms of resistance to realizing the promise of egalitarian Civil Societies. The series will facilitate continued conversation on policy and politics, curriculum and pedagogy, review and reform, and provide a comparative overview of the different conceptions and approaches to citizenship education and democracy around the world.

Titles include:

Zehlia Babaci-Wilhite
LANGUAGE, DEVELOPMENT AID AND HUMAN RIGHTS IN EDUCATION
Curriculum Policies in Africa and Asia

Jason Laker, Kornelija Mrnjaus and Concepción Naval (*editors*)
CITIZENSHIP, DEMOCRACY AND HIGHER EDUCATION IN
EUROPE, CANADA AND THE USA

Jason Laker, Kornelija Mrnjaus and Concepción Naval (*editors*)
CIVIC PEDAGOGIES IN HIGHER EDUCATION
Teaching for Democracy in Europe, Canada and the USA

Marcella Milana and Tom Nesbit
GLOBAL PERSPECTIVES ON ADULT EDUCATION AND LEARNING POLICY

Palgrave Studies in Global Citizenship Education and Democracy
Series Standing Order ISBN 978–1–137–43357–2 Hardback
978–1–137–43358–9 Paperback
(*outside North America only*)

You can receive future titles in this series as they are published by placing a standing order. Please contact your bookseller or, in case of difficulty, write to us at the address below with your name and address, the title of the series and the ISBN quoted above.

Customer Services Department, Macmillan Distribution Ltd, Houndmills, Basingstoke, Hampshire RG21 6XS, England

Language, Development Aid and Human Rights in Education

Curriculum Policies in Africa and Asia

Zehlia Babaci-Wilhite

Graduate School of Education, University of California, Berkeley, USA and Norwegian Centre for Human Rights, University of Oslo, Norway

First published 2015 by
PALGRAVE MACMILLAN

Palgrave Macmillan in the UK is an imprint of Macmillan Publishers Limited, registered in England, company number 785998, of Houndmills, Basingstoke, Hampshire RG21 6XS.

Palgrave Macmillan in the US is a division of St Martin's Press LLC, 175 Fifth Avenue, New York, NY 10010.

Palgrave Macmillan is the global academic imprint of the above companies and has companies and representatives throughout the world.

Palgrave® and Macmillan® are registered trademarks in the United States, the United Kingdom, Europe and other countries.

ISBN 978–1–137–47318–9

This book is printed on paper suitable for recycling and made from fully managed and sustained forest sources. Logging, pulping and manufacturing processes are expected to conform to the environmental regulations of the country of origin.

A catalogue record for this book is available from the British Library.

Library of Congress Cataloging-in-Publication Data
Babaci-Wilhite, Zehlia.
    Language, development aid and human rights in education : curriculum policies in Africa and Asia / Zehlia Babaci-Wilhite.
       pages   cm. — (Palgrave studies in global citizenship education and democracy)
    Summary: "With a Foreword by Martin Carnoy. The debate about languages of instruction in Africa and Asia involves an analysis of both the historical thrust of national government and also development aid policies. Using case studies from Tanzania, Nigeria, South Africa, Rwanda, India, Bangladesh and Malaysia, Zehlia Babaci-Wilhite argues that the colonial legacy is perpetuated when global languages are promoted in education. The use of local languages in instruction not only offers an effective means to contextualize the curriculum and improve student comprehension, but also to achieve quality education and rights in education. Evidence that science literacy is better served through local languages and adapted to local contexts is put forward with a new vision for science learning that invests cutting edge technologies with local context. This vision is crucial to the African and Asian development on their own terms and should take its rightful place as a human right in education" — Provided by publisher.
    ISBN 978–1–137–47318–9 (hardback)
    1. Native language and education—Africa.   2. Native language and education—Asia.   3. Science—Study and teaching—Africa.   4. Science—Study and teaching—Asia.   5. Right to education—Africa.   6. Right to education—Asia.   I. Title.
    LC201.7.A35B33 2015
    372.65—dc23   2015019858

*To the voiceless children deprived of learning in the language
they know best, and thus deprived of their rights to quality
education*

*I hope this book will contribute to the empowerment of all
children*

Contents

Illustrations

Maps

Tables

Figures

Foreword

Africa and Asia were already linked to the Mediterranean and the Middle East by Arab and European traders well before the era of Portuguese exploration began in the 15th century. However, in the 17th and 18th centuries, as European colonization of much of the world brought much of Africa and Asia (and all of the Americas) under European domination to serve European economic and cultural interests, being "part of the world" changed drastically. Africans and Asians became subjugated, dehumanized actors in their own societies.

The massive expansion of the slave trade and the colonization that eventually ensued introduced Africans to a new set of conditions as part of the development of European plantation agriculture and trade with the Americas. When the slave trade ended in the 19th century, the colonizers in some parts of Africa began the "task" of "civilizing" Africans, introducing them to European culture and to agricultural and mining production. This included defining territorial space designed to accommodate competitive European colonial imperial needs (for example, between British Sierra Leone and French Guinea, between British Kenya and German Tanzania, and in South Africa, between competing settler groups). It also included bringing European education to a limited number of Africans, an education that was necessarily in a European language and generally involved converting Africans to a European version of Christianity.

In much of Asia, the process – minus formal slavery – was similar. English, French, Spanish and Dutch colonization installed plantation agriculture (cotton, rubber, tea, palm) in India (including East Bengal, now Bangladesh), Ceylon (now Sri Lanka), Malaya and Indonesia, and Indochina, just to mention the major colonies. Local culture was deprecated and industry (such as cotton textile manufacturing in India) was essentially eliminated by forcing the importation of manufactured goods from the colonizing countries. Again, European education was brought to a limited number of Asians, largely in a European language, and although religious conversion was less of an objective in South Asia and Indonesia, it was important in some colonies, notably the Philippines and Indochina.

It has only been seven decades since the colonies *began* becoming "independent" of European control. Average levels of education in most African and Asian countries have increased substantially and universities have greatly expanded enrollment. However, while the quantity of education has expanded, quality has remained low in Africa and South Asia, and even parts of Southeast Asia, notably Indonesia, constrained by shortages of well-trained teachers, shortages that are exacerbated by the continued dominance of European language instruction in most countries at the secondary and university levels and the large variety of local languages spoken in the territories of many African and Asian nations. In former French colonies in Africa, it is not uncommon in the early grades of primary school to see more than a hundred pupils sitting on an earth floor, being taught French by a single teacher. The pupils, in turn, have no more than an *ardoise* – a small chalkboard – and a piece of chalk to write down the teacher's lessons from the blackboard. In India and Bangladesh schools are also crowded, with teachers often absent, and pupils learn little.

This is not the worst of it. Some areas of Africa (and even Asia, Sri Lanka and Vietnam, Cambodia and Laos, for example) have been racked by years of war or inter-tribal massacres in which children are used as soldiers and others are killed or orphaned by the violence that has marked their lives. There are also waves of malnutrition as drought visits one area, then another. In some regions, such as Southern Africa, the AIDS virus has infected more than a quarter of the population, including many children and teachers.

Nevertheless, with the help of international organizations such as UNESCO and the World Bank, and the many non-governmental organizations – some with private funding, many with individual government funding, and others working as "sub-contractors" for the World Bank – a wide variety of efforts have been sustained over many years in sub-Saharan Africa and South and Southeast Asia to confront health issues and to expand and improve education at all levels.

The "new globalization" in Africa and Asia, at least as expressed in these health and education services, is therefore a complex phenomenon that has simultaneously helped tens of millions of Africans and Asians combat widespread disease associated with common parasites and helped enable access to cheap drugs that fight off the AIDS virus, yet also has continued to "infect" Africa and much of Asia with destructive violence and corruption because of the continued legacy of colonialism.

The efforts to expand and improve education inevitably are part of this complexity. Almost every child now gets a primary education in South Asia and many Eastern and Southern African countries, and overall in India and sub-Saharan Africa more than 70% finish primary school. But the overall gross enrollment rate in secondary school is still low, around 40%, although that proportion is higher in many countries of Asia and some countries of Africa. Yet, in neighboring Sri Lanka and in another former British colony, Malaysia, almost 100% of youth complete primary education, and about 80% finish lower secondary school. There is no doubt that for those who are able to finish secondary school and the smaller group that goes on to university, the economic and social horizon expands. Gradually, as well, more access to education has helped women in Africa and Asia make important gains, and many of the organizations involved in the new globalization have promoted these gains.

This liberating aspect of globalization through health and education initiatives is, however, set in the context of a globalization that has made Africa the scene of violent political struggles to control resources that have relatively little value for Africans. It is set in the context of a globalization that has increased inequality, and that, in turn, has contributed to these violent political struggles. In South and Southeast Asia, inequality is also increasing, and despite quite high rates of economic growth, India and Bangladesh, with their huge populations, are still marked by massive poverty.

The new globalization is not the old colonialism, even though many of its aspects have the distinct look of colonialism. The managers of the new globalization in Africa and Asia are definitely Africans and Asians, many of them developmentalists, many who want higher-quality and more education in their countries. They have the assistance of bilateral and international advisors to help them achieve their goals.

How do the education policies designed to improve literacy and numeracy play into the larger globalization context, including into its most negative aspects of increased inequality and resource and human exploitation? How, in turn, are both international reformers and Africans and Asians – particularly South Asians – caught up in creating educational systems that have difficulty delivering meaningful learning to Indian, Bangladeshi and African pupils, even as they provide greatly increased access to schooling in those systems? These are the great dilemmas of education within the new globalization in Africa and Asia. Good education relies on coherent community and state goals. It relies on states with a firm commitment to develop good teachers – teachers who have those community and state goals at the forefront of their mission to make learning meaningful and exciting to the children and adolescents in their charge. Despite their well-intentioned efforts to create this kind of educational project, international reforming agencies cannot escape educational structures organized to reproduce highly unequal and usually dysfunctional outcomes.

What it takes to emerge from these constraints requires much thought. This book is a step in that direction. Zehlia Babaci-Wilhite examines brilliantly how educational systems and education itself can respond in a more authentic and relevant way to children's learning in the context of the new globalization in Africa and in Asia from the ground up, one issue at a time, reflecting on how different aspects of educational reform in Africa and Asia do not necessarily improve education, and why that may be the case.

Martin Carnoy
Vida Jacks Professor of Education
Stanford Graduate School of Education

Acknowledgments

I am sincerely thankful to Professor Jason A. Laker who inspired me to write a successor to my previous book and who provided great support with excellent feedback throughout. I am also sincerely thankful to my colleagues who graciously welcomed me at the University of California, Berkeley, and provided me with wonderful opportunities to develop this book. I thank my sponsor, Professor P. David Pearson, for his generous time and inclusion, and Professors Jabari Mahiri, G. Ugo Nwokeji and Sam Mchombo for their endless support, which is far beyond what words can express. Thanks to Jacqueline Barber, Traci K. Wierman and Alestra F. Menendez from the Lawrence Hall of Science. It has been a great pleasure to work with you and I am truly appreciative of your efforts to facilitate my work. Thanks also to Professor Macleans A. Geo-JaJa, who mentored my progress during the course of writing several chapters.

I want to acknowledge as well my colleagues and friends in Norway, Tanzania, Nigeria, the USA and all over the world for their support, especially Professors Birgit Brock-Utne, Inga Bostad, Ola Erstad, Robert F. Arnove, Abel Ishumi, Jerome I. Okonkwo, Steve Azaiki, Dr William Bright-Taylor, Azaveli Lwaitama, Mwajuma Vuzo, Jane Bakahwemama, Maryam Ismail, Kimmo Kosonen, Ifeoma Obuasi, Nwobuoka Winifried Obioha Kanu, Julia Obioha and Eleanor Christie.

To the colleagues and students who inspired me at the University of Oslo, the University of Dar-es-Salaam, the State University of Zanzibar, the University of California, Berkeley, the Imo State University, the International Society of Comparative Education, Science and Technology of Nigeria and the University of San Francisco, thank you very much, *asante sana* and *ndeewo*.

I am extremely grateful to the people of Zanzibar, Tanzania and Nigeria who contributed to my work: the headmasters, teachers, lecturers, government officials, students and friends who gratefully gave of their time, facilitated my research during the fieldwork and subsequently corresponded with me via email. A special thanks to Ali

Mwalimu from the University of Zanzibar and Joyce, a wonderful head teacher with excellent staff – *asante sana* for your special attention to my work.

To the many dear colleagues, students, friends, family, thank you very much for all of the genuine support throughout the writing of this book. To my family Kahena J. Wilhite, Alexandre Y. Wilhite, Paul K. Wilhite and Hal L. Wilhite, as well as Louisa Babaci, Hamama Babaci and Fatima Babaci for their support in many different ways, who deserve extraordinary recognition, *merci beaucoup*.

Zehlia Babaci-Wilhite
Berkeley, California
April 2015

Acronyms

ADEA	Association for the Development of Education in Africa
BRICS	Brazil, Russia, India, China and South Africa
CIDA	Canadian International Development Agency
CRC	Committee on the Rights of the Child
DAC	Development Assistance Committee
DFID	Department for International Development
EFA	Education for All
GDP	Gross Domestic Product
IBRD	International Bank for Reconstruction and Development
ICCPR	International Covenant on Civil and Political Rights
ICESCR	International Covenant on Economic, Social and Cultural Rights
ICT	Information Communication and Technology
IMF	International Monetary Fund
Lingua franca	Languages of wider communication, often cross-border languages
LoI	Language of Instruction (synonyms: MoI/MoE)
LOITASA	Language of Instruction in Tanzania and South Africa
MDGs	Millennium Development Goals
MoEVT	Ministry of Education and Vocational Training
MT	Mother Tongue
NGO	Non-Governmental Organization
NOMA	Norad's Programme for Master Studies
Norad	Norwegian Agency for Development Cooperation
NORHED	Norwegian Programme for Capacity Building in Higher Education and Research for Development
NUCOOP	Norwegian University Cooperation Programme for Capacity Development in Sudan
NUFU	Norwegian Programme for Development, Research and Education

ODA	Official Development Assistance
OECD	Organisation for Economic Co-operation and Development
OSC	Orientation to Secondary Class
PITRO	Programme for Institutional Transformation, Research and Outreach
PRIP	Pew Research Internet Project
REDD	Reducing Emissions from Deforestation and Forest Degradation in Developing Countries
SADC	Southern African Development Committee
SAP	Structural Adjustment Programs
SIU	Norwegian Centre for International Collaboration in Higher Education
SYPP	Six-Year Primary Project
TANU	Tanganyika African National Union
UDHR	Universal Declaration of Human Rights
UNDP	United Nations Development Program
UNESCO	United Nations Educational, Scientific and Cultural Organization
UNICEF	United Nations International Children's Emergency Fund
UPE	Universal Primary Education
URT	United Republic of Tanzania
ZEDP	Zanzibar Education Development Plan

Maps

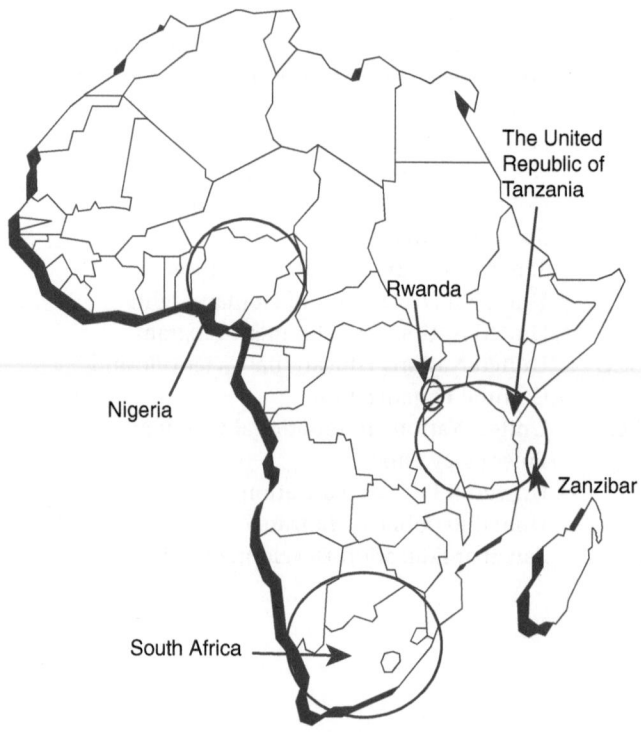

The United
Republic of
Tanzania

Rwanda

Nigeria

Zanzibar

South Africa

Map 1 Africa

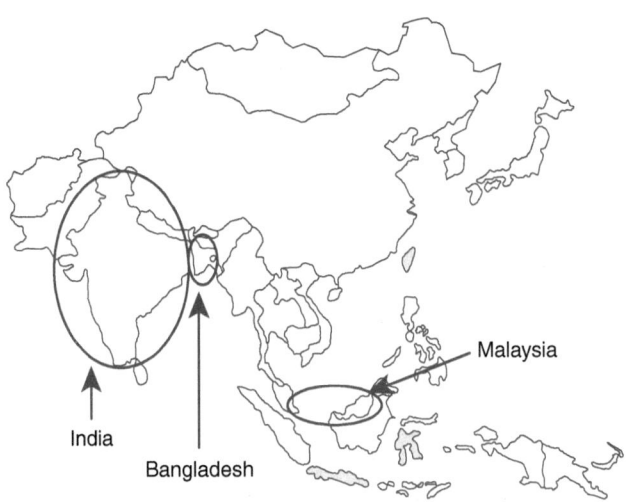

Map 2   Asia

1
Introduction: The Paradigm Shift in Language Choices in Education

In this book I explore the consequences of educational choices for quality education, self-determined development and children's rights in education for democracy in some African and Asian countries. I review theory and educational practices involving the relationship between local knowledge and language of instruction (LoI), and especially their applications to teaching and implications for rights in education. I show that current educational reforms in some African and Asian countries are failing in their efforts to improve education and suggest how educational systems and education itself in these countries could respond to children's learning from the ground up in a more authentic and relevant way. I also reflect on education policies designed to improve literacy and numeracy in the global context and the consequences of a failure of these policies for increased socio-economic inequality. Further, the book addresses how international, Asian and African educational reformers are attempting to deliver meaningful learning to their pupils through expanding access to schooling and confronting the challenges of simultaneously improving student learning. These analyses draw on my extensive work and research experiences on educational systems and reforms in developing countries, both in Africa (the United Republic of Tanzania (URT), South Africa, Rwanda and Nigeria) and in Asia (Malaysia, India and Bangladesh).[1]

The analysis emphasizes that in making English the LoI and ignoring local languages in curriculum change, African and Asian policy makers have been influenced by the still-powerful notion that learning in a Western language will promote development and modernization. This is reinforced by development aid programs directed at the educational sector, many of which encourage and support the use of a non-local LoI. This book will argue that indigenous languages and indigenous knowledge need to be valued in education in order to prepare children to learn and engage in a language they understand best, and as such should be regarded as a right in education.

The book has three main goals. The first is to underline the urgency of acknowledging local knowledge for achieving quality education as a human right. There are an increasing number of books and journal articles that draw attention to the importance of rights to education. These focus on providing universal access to education, but very few have addressed rights in terms of the quality of teaching and learning. My approach is unique in providing a framework for associating human rights with educational choices, emphasizing language choices and local knowledge in educational policies.

My second goal is to flesh out a theory of language policy in education based on a review of theoretical approaches and a comparative study of selected countries (named above) in Africa and Asia. The analysis of educational policy addressed in this book builds on a solid foundation of evidence from Africa and Asia that learning in a local language is critical to quality learning. My third goal is to reflect on the challenges of achieving broader democratization through social justice in education, which involves reforming the politics of globalizing languages and supporting local curricula that emphasize equality and inclusive quality education.

Colonial languages

In the 18th and 19th centuries, colonial linguists and missionaries recorded African languages and classified them as different dialects (Makalela, 2005). In sub-Saharan Africa, African languages were either related to, or were derivatives of, Bantu. The shared communicative base was much broader than that assumed by missionaries and language scientists who have written about African languages from the 1830s to the present, for example the missionaries Isaac Hughes (1789–1870), who transcribed seTswana, and Andrew Spaarman (1747–1820), who transcribed isiXhosa in South Africa. These transcriptions led to the linguistic separation of these closely related languages. Leketi Makalela (2005, p. 151) calls this process the "de-Africanisation through displacement of African languages". Furthermore he argues that "African languages are not so different as to impede communication, as it is canonically assumed" (2005, p. 166) and that a harmonization, rather than a segmentation, of languages ought to be made. However, he notes that:

> The notion of harmonisation is often misinterpreted to mean that some African languages will be killed and that people will lose their languages and identities ... But as Webb and Kembo-Sure (2000) rightly put it, this process of harmonisation does not take anything away from the speakers, but rather adds ... a core written Standard for literacy, which learners from different languages acquire at school while retaining their home or spoken varieties.
>
> (2005, p. 168)

If local languages were harmonized, it would help to protect traditions through stories, myths and songs (Babaci-Wilhite et al., 2012a). Languages with a colonial legacy, such as English, French, Portuguese and Spanish, to a smaller extent, continue to be used as official languages in many developing countries today. Africans were forced to use European languages, and this constituted a form for colonization of the mind (Ngugi wa Thiong'o, 1994). This same mental colonization took place in several Asian countries. In British Colonial India, English was made the LoI by the English Education Act of 1835, but only with the purpose of generating a cohort of clerks, peons and petty functionaries to oil the machinery of empire (Viswanathan, 1989). Thus, ironically, both English and the mother tongue (MT) can play a divisive role in gatekeeping linguistic capital (Hornberger & Vaish 2009, p. 308).

Globalizing English

English is a particularly powerful globalizing language that is influencing debates on choice of LoI in many developing countries in Africa including India, Bangladesh and Malaysia, which will be elaborated in Chapters 2 and 6. Ayo Bamgbose (2003, p. 421) notes that "language has a pecking order and English has the sharpest beak". It carries with it a cultural context foreign to the local contexts for education (ibid.).

The embedding of English in education is a method for creating dependency, accomplished through the institution of European languages, metaphors and curricula. In recent years, the use of English as a LoI in postcolonial countries has been a subject of debate and research. Many scholars argue that English intervention in learning promotes and prolongs neocolonialism and that its expansion should be halted (Phillipson, 1992; Mazrui, 2003; Alidou & Brock-Utne, 2011; Skutnabb-Kangas & Heugh, 2012).

Today, English is used as an official or semi-official language in over 60 countries in the world and has a prominent place in a further 20 (Majhanovich, 2014). Globally, it is the main language of books, newspapers, airports and air-traffic control, international business and academic conferences, science, technology, medicine, diplomacy, sports, international competitions, pop music and advertising (Mazrui, 1997).

Braj Kachru (1990) describes the spread of English as three concentric circles to explain how the language has been acquired and how it is used. The first inner circle represents its use as an MT or/and a first language. The second circle comprises countries colonized by Britain where non-native speakers learn English as a second language in a multilingual setting, as is the case in Tanzania. The outermost circle consists of countries that dedicate several years in primary and secondary education to the teaching of English as a foreign language, such as most European countries.

Some scholars, namely Michael Kadeghe (2003), regard English as a valuable asset for global business and cross-cultural communication. Many language policy makers have adopted this view particularly in wealthy nations like the United States of America (USA) and the United Kingdom (UK), where large amounts of "foreign aid" moneys are spent on promoting English, and in sub-Saharan Africa where English is now often the sole official LoI at all levels of education (Mazrui, 2003). These perspectives ignore the issues of quality learning and cultural identity. Robert Phillipson (2000) argues that this increasing global influence of English constitutes a linguistic imperialism and counterposes the preservation of native languages in line with Tove Skutnabb-Kangas's (2000) work on language as a human right.

Virtually all African linguists and educationalists argue for the advantages of the use of an African language as the LoI (Fafunwa,

1990; Qorro, 2003; Prah, 2009; Okonkwo, 2014). Children taught in any of the language varieties similar to their MT will have better learning comprehension than those taught in an adopted foreign language such as English, and furthermore MT education leads to more effective teaching of sciences and mathematics. An effective language policy takes care that the languages taught in education reflect everyday communication patterns.

Carol Benson and Kimmo Kosonen (2013) conclude from their research that learning in one's MT allows for better learning of all subjects, including the learning of a second language. The language that a child masters best is the language used at home and in the local surroundings; however, the choice of language for a local school is complicated by the fact that in many African contexts there are several languages used in the community. There is not always an obvious choice of local language and this has led to many local debates on whether one of the local languages should be used or, as in the case of East Africa, whether a pan-African language such as Kiswahili should be used as a LoI (Brock-Utne, 2007; Vuzo, 2009; Babaci-Wilhite, 2013a).[2]

The cost of using multiple MTs in different regions is high, and there are also debates on whether this separation into regional LoI is feasible. I acknowledge the importance of this debate and the difficulties involved in the choice of a local or pan-African language, but I derive from the literature that due to the fluency of some African languages, such as Kiswahili in Zanzibar and because Kiswahili is a locally constructed language that is related to the vast majority of East African languages, that it is an obvious choice for primary schooling in Zanzibar.

An important issue in choice of a local LoI such as Kiswahili is its reinforcement of local identity. Identity is strongly connected to parents' beliefs, to the language spoken at home and to local culture. The overwhelming message from research in Africa is that using a language that learners use in their everyday lives will improve learning and help to maintain the connection to the local cultural context. The use of a local language in education will contribute to literacy and strengthen cultural identity. According to Moshi Kimizi (2012), the use of a local language as a teaching medium will also affect a child's self-esteem. The learning process can be done effectively only if a child feels that her or his identity is acknowledged. The

Figure 1.1   Children's punishment when they speak their language in school[3]

best learning environment will be created when a child feels that their language has value. Should the local language be rejected, this is equivalent to the rejection of local identity. Research shows that this sense of rejection is affecting children's sense of identity in several African countries (Mazrui, 2003), which is a violation of human rights (Figure 1.1).

Many schools around the world use English as the LoI in the expectation that it will bring greater academic success and mobility for their students. Several scholars in the study directed by Adama Ouane and Cristine Glanz (2006) leave no doubt that the use of MTs facilitates the learning processes in schools. Martha Qorro (2003) confirms that there is a belief in Tanzania that learning in English will improve the learning of the language; however, she points out that the LoI has another important function, in that concepts are communicated to children in the language they understand best. Bamgbose's (1984) study in Nigeria and Grace Bunyi's (1999) study in Kenya confirm this point, showing that when science instruction was conducted primarily in English as opposed to a native tongue, students were unable to apply concepts they had learned in class to practical situations at home. Sunil Loona (1996, p. 3) reinforces the point on the power of local language to communicate concepts when he writes, "Learning a second language does not imply the development of a totally new perspective, but rather the expansion of perspectives that children already possess." It is important to make the points that learning in a language and learning a language have two different functions, and to combine these functions will slow and possibly stop the process of learning (Qorro, 2004). This difference has to be understood and acknowledged in the curriculum.

In Africa and several Asian countries, the policy of switching from a local language to English midway through the schooling process gives the impression that African and Asian languages are inferior to English and that the local language is somehow inadequate for engaging with complex concepts. This reinforces the sense of inferiority of local culture and at the same time is disadvantageous for those who have had little exposure to English at home. This choice in Africa and Asia has contributed to the formation of a national identity and cultural identity as exemplified in Tanzania, India and South Africa. Whichever context a child is in, s/he can hardly achieve quality learning when there are identity problems. Moreover Kosonen (2010) argues that the use of MTs produces better learning results in all school subjects and also reduces repetition and drop-out.

Local languages as cultural capital

The use of a local language in the educational system also contributes to self-respect and to pride in local culture. By reinforcing the importance of local languages, one reinforces the interest in local knowledge and culture. Ideally, one would choose a non-dominant local LoI, but in cases in which this is expensive and practically difficult to implement a local language, such as Kiswahili, that has local roots and is widely used in public spaces is a good second choice (Babaci-Wilhite, 2010).

I agree with Birgit Brock-Utne (2007) who is critical of the consequences for the development of self-respect and identity of not using the language one normally communicates in as a LoI in school. Language is part of one's identity and part of one's culture and a local language should be used to reinforce both through schooling. This is consistent with Rwantabagu's (2011, p. 472) argument based on a study in Burundi where he argues that "The possession of one's language and culture is at the same time a right and a privilege and members of the younger generation should not be denied their native rights and advantages, as provided for in article 27 of the Universal Declaration of Human Rights" (UN, 1948). Furthermore, he points out that after a series of education reforms, there is still a debate in Burundi over the use of French or Kirundi as LoI. He argues for the importance of local LoI as a basis for better knowledge acquisition, and that policy makers should take evidence-based research on language and education into consideration. He also points out that African languages will enable African cultures to exist and will ensure the survival of African languages, referring to Senghor's (1976, p. 10) point that culture evolves which entails that our school systems should aim both at cultural authenticity and openness to foreign influences, said to be "the humanism of the new millennium" (ibid.). Furthermore, he concluded that within the context of globalization, giving prominence to African languages and cultural values could be based on "interdependence and complementarity between cultures and nations", addressing one of the most important challenges of the 21st century.

In sum, the implications of education in Africa and Asia for identity and development argue for the choice of a local LoI in education. Education must be centripetally oriented, and based on the principles of respect for human rights and cultural dignity. It must give consideration to local realities and direct its intellectual efforts and curriculum toward the achievement of freedoms that are consistent with education as a human right (Babaci-Wilhite & Geo-JaJa, 2011).

According to Skutnabb-Kangas (2000), the most important linguistic human right in education for indigenous peoples and minorities, if they want to reproduce themselves as peoples/minorities, is an unconditional right to MT-medium education in non-fee state schools, exemplified by a policy put into place by Imo State in Nigeria in 2014. Moreover, she argues that binding educational Linguistic Human Rights are more or less non-existent in African countries (ADEA, 2001) and states a pessimistic but realistic estimate that 90–95% of today's spoken languages may be very seriously endangered or extinct by the year 2100. This means another round of colonization of the mind through cultural assimilation of non-local language and culture. Since much of the knowledge about how to maintain the world's biodiversity is encoded in the small indigenous and local languages, with the disappearance of the languages crucial ecosystem knowledge will also disappear and be replaced with "Western" scientific knowledge if we do not acknowledge local LoI as important in education.

Local languages improve teaching and learning, as well as having significant subsidiary benefits for the society. The results from many studies around the globe demonstrate that English as a LoI hinders educational development and reinforces social inequality. A result of the transition to English LoI is a diminution of "cultural capital" in poor and socially excluded groups (Bourdieu, 1977). According to Pierre Bourdieu (ibid.), the ability to function well in school and in society will be dependent on certain surrounding factors such as parental education, the number of books in the home, the amount that a child is read to and the amount that a child is talked to. Loona (1996, p. 6) writes, "Children do not arrive at school with equal amounts of knowledge of the world...Differences in experiences in homes and in their daily lives can lead to some children having lesser or greater amounts of knowledge in some knowledge-domains than other children." Language is used in the learning process inside and outside of the house. Children of elite parents are more likely to have access to English literature and films, and to have travelled, and thus to have been exposed to the use of English in different contexts. Therefore the use of English as LoI gives advantages to elite families and reinforces disadvantages for others. In effect, the skewed cultural capital will be reinforced and institutionalized in the education system (Bourdieu, 1977; Loona, 1996).

As discussed above, cultural context is crucial for learning; however, current classroom education does not take advantage of the immense learning opportunities available at home, in communities and in workplaces (Samoff, 1999; Erstad et al., 2009; Geo-JaJa & Azaiki, 2010; Babaci-Wilhite et al., 2012). Martin Carnoy (2007, p. 95) argues, "How much pupils learn in school depends greatly on what concepts they are exposed to, how much time they spend studying these concepts, and how effective their teachers are in communicating them." A review undertaken by a joint research team from UNESCO and the Association for the Development of Education in Africa (ADEA) concluded that the interconnectedness between language, communication and effective teaching and learning is generally misunderstood outside expert circles (Ouane & Glanz, 2006). Using a foreign language as a LoI makes the language a barrier rather than an aid for both teachers and students.

James Cooke and Eddie Williams (2002) cite numerous studies (Machingaidze et al., 1998; Nassor & Mohammed, 1998; Nkamba & Kanyika, 1998) conducted in several African countries to show that "the vast majority of primary school pupils cannot read adequately in English, the sole official language of instruction" (p. 307). Cooke and Williams (2002) go on to state, "If children in developing countries have little exposure to the LoI outside the school, and if teaching the LoI is ineffective inside the school, then low-quality education is inevitable" (p. 313). As the majority of these students leave school with no literacy and a low level of competence in a language they use very little outside the classroom, to propose that receiving their education in English disadvantages them is a severe understatement.

English language education is put further into question when examining the inequality it perpetuates between its immediate bene-factors, the relatively wealthy, and those for whom it has no practical use, the severely impoverished. In addition to possessing the means to access larger markets and coveted white-collar jobs, the relatively wealthy urban groups also have better educational opportunities leading to greater levels of English proficiency than the more dis-advantaged urban and rural poor are able to acquire. English then becomes an upper-class language, which the poor hold in great esteem but cannot effectively access because of the low quality of their education and their disadvantaged economic status.

Learning to read and write in a local language is directly corre-lated with the improvement of a student's abilities to think critically about their own conditions and about the world. Using local lan-guages as LoI provides a sustainable benefit in the form of national cohesiveness for nation-building and cultural identity. Children of all backgrounds will be able to perform better in school with local languages. This path forward will contribute to our understanding of quality education and children's confidence in their community as well as social equality.

Organization of the book

In this introduction I have elaborated the book's rationale and pur-pose and introduced the three domains that will be explored in the book: global versus local languages in schooling for appropriate development; development aid and collaboration; and human rights in education for quality learning and teaching. In general terms, quality education corresponds to basic education as set out in the World Declaration on Education for All (EFA), but must also ensure human rights through localizing education in local language and context. It should take account of the educational, cultural and social background of the students concerned. It demands flexible curricula and varied delivery systems to respond to opportunities of communities and the needs of students in different social and cultural settings.

Chapter 2 reviews the literature and provides a theoretical framework for assessing the choice of local LoI in a human rights perspective. The central theoretical focus in the chapter will be on the explanatory power of theories related to development, modernization and human rights to explore the importance of the choice of LoI and its consequences. The theoretical framework laid out in the chapter incorporates the importance of local context, using a local language and emphasizing the development of local capacity on local terms. Such a multiple approach emphasizes the importance of indigenous concepts articulated in their natural environment. Any local curriculum that ignores local languages and contexts risks a loss of learning quality and a violation of children's rights in education.

Chapter 3 explores the place of development aid in education, focusing on Africa and Asia. It traces the historical trajectory of development aid from an original focus on narrow economic concerns during the immediate post-World War II years to the current focus that incorporates social issues and challenges to education rights. To sustain and deepen this focus on education rights and to make development truly meaningful to recipients of aid, serious commitment on the part of citizens, institutions and donor communities toward development collaboration is imperative.

Chapter 4 explores human rights in development aid and the role of human rights in a holistic approach to education. The premise is that the Rights-Based Approach addresses the issue of recipient ownership and promotes development on local terms in Africa and Asia. This approach acknowledges that recipients have rights in education and suggests that LoI as a cultural and human right in schooling should be central to development aid.

Chapter 5 describes and assesses initiatives by the Norwegian Agency for Development Cooperation (Norad) to improve and expand collaboration with universities in developing nations through the sponsorship of partnerships between Norwegian universities and universities in the South. This chapter uses as a case study a collaboration between the University of Oslo and the University of Dar-es-Salaam in Tanzania from 2009 entitled "Programme for Institutional Transformation, Research and Outreach" (PITRO) to acquire the necessary knowledge and skills both for increasing economic productivity and for reducing poverty.

Chapter 6 demonstrates the links between human rights, language choice for appropriate development and social justice. Using a local language satisfies the rights criteria of availability, accessibility, acceptability and adaptability. These should be common to education in all its forms and at all levels. In many African and Asian countries, the goal of rights to education is becoming increasingly remote, let alone that of rights in education. With this understanding and with the awareness of the education challenges of nations and millions of people throughout Africa and Asia, rights in education remains a distant goal. It is imperative to overcome these obstacles in order to bring social justice and appropriate development.

Chapter 7 builds on a solid foundation of evidence from Africa and Asia that learning in an indigenous language is critical to quality learning and important in the reinforcement of cultural identity. In this chapter, I review and assess the literature on language choices in education, teaching and quality learning, education for self-reliance, language identity and power. I compare the recent changes in LoI in Zanzibar and Malaysia in light of the results of fieldwork in mainland Tanzania and Zanzibar as well as a review of the literature in Malaysia and Nigeria. I present the theoretical framework for analysis of why policy makers ignored this evidence on the relationship between indigenous languages and quality learning for all.

Chapter 8 reviews and assesses the debates on LoI and their consequences for quality of learning. The scope is global, but it gives special attention to a comparative study between India and South Africa, Bangladesh and Rwanda. I compare official documents on the debates on choice of LoI in South Africa, India, Rwanda and Bangladesh. The cross-national comparison allows for an analysis of why some Asian and African countries have continued using colonial languages in their educational systems, while others have chosen a local LoI. Experiences from some African and Asian countries show clearly that contextualized education will contribute to their transformational educational objectives and foster economic growth and inclusive development. I examine the implications of the choice of LoI for work prospects and participation in the global economy.

Chapter 9 addresses why many educational practitioners and policy makers continue to ignore language and culture as a central ingredient in science literacy and the consequences for learning science. Past projects in Africa, and especially in Nigeria, have shown how literacy and science can be better achieved in a local language, yet Zanzibar has recently changed its LoI from a local language to English for science and mathematics.

Chapter 10 concludes that education is critical to development and that both education and development should accommodate the broader context of human rights. By drawing out the distinction between rights to education versus rights in education, I conclude that quality education is the key to success in to and for rights in education. This conclusion reiterates the argument that the use of a local LoI is indispensable for inclusive development and that linguistic and cultural rights should be integral to education systems, as they are critical to social justice in a democracy.

Furthermore, I briefly address recommendations on how to work toward the achievement of two important goals for education in developing countries, the integration of global and local learning in the digital age and the retention of local LoI in the digital age. To finalize, I argue that the integration of E-learning into conventional curricula, using a local language and local knowledge, ought to be defined as a global justice tool for inclusive development.

2
Educational Issues in Africa and Asia

This chapter reviews theories appropriate to analyzing the relationship between language choices in education, learning and development in Africa and Asia and addresses the question of why local languages are not used as LoI in most African and some Asian countries. It proposes that an important explanatory factor is the legacy of dependency relationships between Western, developed countries and many countries in the periphery of the post-World War II development axis. Despite sporadic resistance, including the policies of Nyerere in Tanzania in the 1960s and 1970s, the perception in Africa that development is synonymous with emulating Western institutions continues to have relevance for African and Asian education policy in many countries. This creates fertile ground for the extension of global markets for English language curricula and support materials. It will be argued that the theories of Paulo Freire and Amartya Sen could provide a framework for transforming education systems in Africa and Asia so that they are reoriented to the development of local capabilities and the reinforcement of local cultures and languages.

Dependency in language imperialism

Colonial and post-World War II development have promulgated an educational framework for many African countries in which a language of the metropolitan center of the world-system, English, has become the LoI in education. In effect, the colonial mind embedded in African education has been perpetuated under the guise of international development. Education has been an export commodity from the center and is accepted in Africa as a result of Western-based ideas of what it means to develop. Dependency theory provides an explanation for this acceptance of Western ideas and practices, and has been important to understanding why African and some Asian countries are still using English as LoI.

Economists Raul Prebisch and Hans Singer were among the founders of dependency theory in the late 1950s, under the guidance of the Director of the United Nations Economic Commission for Latin America (Ferraro, 1996). They argued that unbalanced economic power put many developing countries at a disadvantage in their interactions with developed countries (ibid.). Dependency theory and the world-systems analysis with macro-scale approaches to social analysis and social change, developed, among others, by scholars such as Andre Gunder Frank (1966) and Immanuel Wallerstein (1974), is defined by Robert Arnove as

> a protest against neglected issues and deceptive epistemologies ... a call for intellectual change, indeed for 'unthinking' the premises of nineteenth-century social Science ... an intellectual task that has to be a political task as well, because ... the search for the true and the search for the good is but a single quest.
>
> (2009, p. 115)

Frank (1966), one of the earliest dependency theorists, argued that an adequate theory of development is not possible without accounting for how past economic and social history gave rise to present underdevelopment. Furthermore, Frank (1966, p. 27) claimed that "our ignorance of the underdeveloped countries' history leads us to assume that their past and indeed their present resembles earlier stages of the history of the now developed countries". For the dependency theorists, underdevelopment is a wholly negative condition, which offers no possibility of sustained and autonomous economic activity in a dependent state. Furthermore, he argues that "economic development was neither self-generating nor self-perpetuating" (1966, p. 27) in Africa. In line with Frank, Wallerstein (1974, 1980) elaborated dependency theory into a global system through his world-systems analysis. Wallerstein was engaged in an attempt to explain how development through global power structures favored the developed centers and why developing countries accepted prescriptions for development that lead to dependent relationships. Wallerstein's analysis provided a model for understanding the change in the interlocking global system after War World II. He writes of how rich nations of the world – core countries – have forced disadvantageous trade relationships on the poor – the periphery – in what he refers to as "unequal exchange", and points out that these have been mainly advantageous for the former. Furthermore, Robert F. Arnove (1980) has applied this model to education and questioned the imperialist approach within the universal education system, which divides the center from the peripheries by using English in teaching and learning for the privilege of only a few.

Johan Galtung (1971), Samir Amin (1998), J. Ngugi wa Thiong'o (1993) and Alamin Mazrui (1997) have all contributed to the development of theory relating imperialism and power imbalances in global relationships to dependency and underdevelopment. Galtung (1971) argued that socio-economic developments in the nations of the periphery (developing countries) were heavily affected by their dependent relationship with the USA and the UK. Alastair Pennycook (1994) analyzes the spread of English using Galtung's (1971) concept of center and periphery. He makes the point that those who are in power in the periphery have strong links with the center, since most of them have either been educated in the countries of the center or through one of the languages of the center.

Applying dependency theory to African and select Asian countries, the colonial and post-War development agenda as promulgated a framework of dependent education in which English has infiltrated African educational systems. After decades of exposure to the fallacies of this acquiescence to globalizing English, many developing countries have made efforts to opt out of dependent relationships. However, the notions of progress, advancement and development that characterized development theory and practices continue to influence African political choices, including those made in the domain of language choice for education. Furthermore, as Cooke and Williams (2002) write, "Far from being a source of unity, the use of English in education in many poor countries has become a source of national disunity" (p. 314) and "the use of English to achieve development has significantly contributed to the socioeconomic and political instability of most African countries" (ibid., p. 315). Thus, the popular notion that the promotion of English as the official language of education creates national unity in developing countries is an illusion. While English may indeed have a beneficial role to play in the overall economic and societal development of developing countries, it should not be assumed that English is the solution to any developmental needs. Nor should it be promoted at the expense of much-needed language education at the local level (Bruthiaux, 2002).

Language dependency in education has been perpetuated through development aid subsidies and donations (Samoff, 1999; Brock-Utne, 2000), offered in the form of scholarships, courses, training, as well as through research cooperation, promotion of textbooks, gifts of books, supplying of teachers and supplying experts in curriculum development and advisory work. Several examples illustrate how the World Bank's financial support has been directed toward consolidating European languages in Africa (Mazrui, 1997; Brock-Utne, 2000; Prah, 2003). While the UK and the USA have provided gifts of books and experts in curriculum, this type of aid paves the way for intellectual and economic dependence (Buchert, 1994; Mazrui, 2003). According to Mazrui (1997, p. 45), "the World Bank and the IMF naturally have a vested interest in this interplay between linguistics and economics". Mazrui (1997) argues that "imperialist control can also be approached from the point of view of language, not as a reservoir of culturally-bound world views, but as an instrument for the communication of ideas". The World Bank must acknowledge and account for the failures of development of the post-World War II period. Yash Tandon's book on aid dependency claims that the International Monetary Fund (IMF) has been equally ineffective and "most developing countries do not need it any more" (2008, p. 121). He goes on to claim that IMF credibility is at its lowest point and that it has "little expertise" as a development institution (2008, p. 121). He argues that for development to be successful, it must be pursued according to the terms of the developing country. Development is

... self-defined; it cannot be defined by outsiders ... development is a process of self-empowerment ... development is a long process of struggle for liberation from structures of domination and control, including mental constructs and the use of language.

(ibid., pp. 12–13)

This is in opposition to the Western concept of development, which is generally seen as a process of changing or converting institutions and values in developing countries to match those of the West. Furthermore, Yash Tandon (2008) claims that it is important to distill from the imperially imposed system of values those which are indeed universal and recognized in all cultures and civilizations, though in different forms and languages. There are aspects of cultural values that acquire universal validity and recognition through multicultural interaction and mutual learning in which no culture is superior to others. The imperial project that seeks to impose nationally or regionally specific values, and the intent of the donors to serve imperial interests, must be distinguished from this broader historical vision. There are good reasons for questioning the true interests and intentions of developed countries involved in development aid. In line with Tandon's arguments, development is Africa's responsibility and not that of the donors. A new mindset is needed in which local knowledge and local languages are privileged in development projects.

The changes forced on Africa in the 1980s had dramatic consequences for African countries. Tanzania's Education and Training Policy in 1995 stipulated that all levels of education are open to private actors. Brock-Utne notes that:

> From that date, the increase in English-medium primary schools has been spectacular... Tanzania was forced by the World Bank and the International Monetary Fund into structural adjustment measures like cutting down on public expenditures, including the education sector, the opening up of private schools and the liberalization of the text-book market.
>
> (2005, p. 73)

Critics of development aid emphasize the importance of rethinking development and changing educational policies in order to take account of local curriculum with local values, local culture and local languages. This argument is highly relevant for explaining, and restraining, globalizing ideas about education. Furthermore, Arnove argues for the need "to take up world-systems analysis as the necessary framework for understanding educational trends around the world, from curriculum reform to the LoI and the outcomes of school expansion" (Arnove, 2009, p. 102).

Globalization and commodification of education

New globalization and the commodification of education (Carnoy, 1999; Geo-JaJa & Yang, 2003; Arnove, 2012; Klees, 2013; Amin, 2014) were readily accepted in Africa due to embedded local associations between development and inculcation of Western ideas of education and development. As intimated above, the unequal exchange postulated by dependency and world-systems theories for commodities and trade relationships is highly relevant for understanding the Western influence on African languages and education.

Frank's (1966) observation is highly relevant to the language debate: generations of scholars from developing countries "face the task of changing this no longer acceptable process of eliminating the miserable reality...by importing sterile stereotypes from the metropolis which do not correspond to their satellite economic reality and do not respond to their liberating political needs" (p. 37). The export of English as LoI is a classic example of a sterile stereotype that does not respond to local needs.

Mazrui (2002) argues that English and some other European languages in use in Africa continue to mesmerize African policy makers. The consequence is that learning continues to be based on Western concepts and thinking, which reinforces the cycle of dependency. Thus, almost all educational policy formulations, including the role of language in education policies in Africa, emanate from the West. Learning continues to be based on Western concepts and thinking, which reinforces dependency (Mazrui, 2003). English in education is associated with modernization by both policy makers and the parents of school-aged children. Policy makers rationalize the change of LoI by arguing that they are acquiescing to the demands of parents, who believe that learning in English will equip their children with the capacity to find work in the globalized economy. My research in Zanzibar reveals that when it comes to choice of LoI, myths such as this one affect both parents and educationalists. They foster the misunderstanding that learning in English will improve English skills without having negative consequences for learning capacity. The work of Zaline Roy-Campbell and Qorro (1997) shows that if the LoI is not a language used in everyday life, the ability to develop new perspectives will be inhibited (Babaci-Wilhite, 2013a).

Another globalizing trend is the marketization of education, manifested in various ways. According to Arnove (2003, 2012), the importation of Western languages and curricula emphasizes a homogeneous national curriculum and high-stakes standardized tests that fail to account for the variety of abilities and talents students possess. The use of English as LoI could also lead to an increase in after-school tutoring, which is costly and only affordable to middle and upper income families. According to Bourdieu (1977, p. 89), "the unequal likelihood of extra-curricular work among the different categories of students" can explain why some students do better than others in schoolwork. David P. Baker and Gerald K. LeTendre (2005) discuss the emergence of private tutoring, known as "shadow education", which improves children's performance within schools, increases academics' competitiveness, diversifies education opportunities, opens new avenues for peer socialization and channels private funding to public schools, which favor only a few. Children whose parents can afford to pay are more likely to be successful in school and to be rewarded by performing better on tests and other forms of evaluation. Iveta Silova (2010) points to the increasing inequalities resulting from the hidden privatization of public education on children, families and societies (see also Zajda & Geo-JaJa, 2010). Mori and Baker (2010, p. 46) argue that "shadow education can be shown to corrupt the quality and goals of public schooling" and warns that policy makers and education analysts "can easily go astray in interpreting trends in schools... from a purely local or national perspective" (Baker, 2009, p. 961).

Consequently, what is often missed in policy makers' calculations is the common and deeper cultural forces at work behind the peculiar characteristic of a specific problem at a particular time and place. There are a number of cases where the ignorance of these deeper forces has led to surprisingly unsuccessful policy implementation. Embedded in these ideas of educational progress are market principles such as efficiency, competition and quantification which Carnoy (1999), Joel Samoff (1999), Macleans A. Geo-JaJa (2009) and Amin (2014) write are characteristics of the commodification of education. Grazia Scoppio (2002) broadens this "marketization" to embrace the assumption that customers, meaning the students and their parents, are the best judges of the value of services rendered and that they should be given the choice among competing institutions (in areas where there are more than one institution). English as LoI is commodified and sold as a valuable tool for achieving economic success; however, in most cases this expectation of added value is an illusion – a myth (Qorro, 2005). Furthermore, she argues that even with an education, very few people in these societies have the means to access the white-collar jobs or larger markets that require any knowledge of English at all. The majority of the poor instead participate in their local, informal economies, which can involve 50% of the labor force and account for 40% of the Gross Domestic Product (GDP) in underdeveloped countries (Bruthiaux, 2002). English is virtually unused in this massive informal sector, diminishing the economic value of English as LoI.

These ideas of English supremacy and commodification of education are transmitted through development programs and through other forms of global interaction. Mazrui (2002) and Samoff (2007) demonstrate that intellectual and scholarly dependency in Africa still exists.

Education for self-reliance and empowerment as capability approach

Theoretical perspectives on learning and the culture-enabling role of education proposed by Julius K. Nyerere (1968), Paulo Freire (1977) and Amartya Sen (1999) are important to an analysis of language in learning debates. Each of these scholars emphasizes in their own way the importance of drawing on local knowledge and local culture in both education and development strategies.

The first President of Tanzania, Julius K. Nyerere (his presidency extended from 1962 to 1985), provided an alternative to the modernist vision for development in his Arusha Declaration of 1967. Nyerere's theory of self-reliance was developed and applied in Tanzania in the 1960s and 1970s. As outlined above, his theory no longer governs educational policy in Tanzania, but I argue nevertheless that it has relevance today in Tanzania's efforts to achieve equal access and fairness in education. Nyerere's ideas on education for self-reliance for Tanzania not only provided a basis for true Tanzanian development, but also provided a model for other countries in the

South. Nyerere was an educational visionary who insisted on a rethinking of the relationship between general education and formal schooling. Nyerere wrote that:

> We have not until now questioned the basic system of education which we took over at the time of Independence. We have never done that because we have never thought about education except in terms of obtaining teachers, engineers, administrators, etc. Individually and collectively we have in practice thought of education as training for the skills required to earn high salaries in the modern sector of our economy.
>
> (Nyerere, 1968, p. 267)

His vision of integrating local development and local education was seen as a way of resolving many of the problems of colonization and one-way development. Designing education in a way that accounts for local culture, language and livelihoods would also bring autonomy and pride in the country. Nyerere (1968) was clear on the point that one country should not depend on another to educate its citizens. He advocated that the developing countries of the world should build their own school curricula. This freedom for Tanzania to define its own educational philosophy and system would help Tanzanians to achieve respect and freedom from repression. This would also inspire local pride and cultural learning. He wrote, "Colonial education in this country was therefore not transmitting the values and the knowledge of Tanzania society from one generation to the next" (1968, p. 47). To motivate the active mind, one has to take into consideration the variations in different societies, differences in knowledge and different ways of teaching to achieve quality education, in which language plays a crucial role. The knowledge one learns in school is only one contribution to a complete education. Equally important is that an educational curriculum incorporates the country's own values and traditions. In historical and contemporary Africa, education is not seen as directed toward an economic outcome, but is viewed as a holistic experience that produces both individual and social persons.

Freire's (1970) theory on pedagogy is very relevant to a much-needed transformation today in education. He raises questions about formal versus informal learning and the role of schooling in education. Furthermore, Freire's theories articulate the intimate relationship between education and development, particularly the connection between individual empowerment and democratic ideals, where people use their education to critically analyze and change their conditions. Arnove et al. (2003, p. 329) suggest that Freire's ideas on literacy campaigns and popular education programs be put in place in order to "(1) increase access to schooling, (2) democratize school administration, (3) improve instructional quality, (4) expand educational opportunity for working youths and adults, and (5) contribute to the formation of critical and responsible citizens" (2003, p. 329). Arnove et al.'s (ibid.) overall assessment, in line with that of Lindquist Wong (1995), is that the model of educational reform during Freire's administration was one of the most successful in terms of its process and outcomes because the state and the civil society worked in tandem rather than in opposition. Ira Shor and Freire (1987, p. 2) formulated Freire's conceptualization of education thus: "teaching in a classroom is a very practical activity, even though everything touched on in the classroom is the tip of a theoretical iceberg". The curriculum should aim at furthering local knowledge and reducing the hierarchical nature of teaching. If true openness to knowledge is to be achieved, the idea that the teacher knows best and the submissive student must be abandoned. Freire further argues that:

> The teacher talks about reality as if it was motionless, static, compartmentalized, and predictable ... his task is to "fill" the students with the contents of his narration totality that engendered them and could give them significance. Words are emptied of their concreteness and become a hollow, and alienated verbosity.
>
> (1993, p. 52)

It is important to impart knowledge in ways that inspire independent thought and empower students. Having students only memorize information by heart in a foreign language should not be the basis for a curriculum. As formulated by Freire (1993, p. 60), "Liberating education consists in acts of cognition, not transferals (sic) of information". Every society should liberate its educational system, not by transferring knowledge but by inspiring people to reflect on how to achieve a better life. Even though education encompasses much more than schooling, schooling is nonetheless central to knowledge acquisition, and primary schooling was formally accepted as a human right more than 50 years ago (Colclough, 1993). The lives of adults and children everywhere revolve around the school. Noel F. McGinn (1997) argues that teachers and principals should be given more control in curriculum development since learning takes place in schools. This is not the case in many African countries. In most cases "Knowledge is produced in a place far from the students" (Shor & Freire, 1987, p. 8) and students are used to the transfer of knowledge with an "official curriculum that asks them to submit to texts, lecture, and tests, to habituate themselves to submitting to authority" (Shor & Freire, 1987, p. 11). Non-local curricula tend to have similar content and structure. According to Samoff (2007, p. 60), "effective education reform requires agendas and initiatives with strong local roots". In studying the ways a curriculum engages with global and local knowledge, John W. Meyer's (1992) argument is relevant: an educational curriculum is a global, modernizing institution which expresses an increasingly global culture that is largely independent of national policy. Along the same lines, Michael Young (2008) argues that all too often, curriculum is responding to external political and economic forces rather than to the internal conditions for knowledge acquisition. These theories of Freire, Nyerere, Meyer and Young address "what kinds of knowledge should be the basis of the curriculum and how they can be made accessible to the majority of students" (Young, 2008, p. 11). These perspectives were important in analyzing changes in teaching and learning in Zanzibar. Every continent seems to adapt their education to their cultural context – except Africa, which is still using European languages and European curricula in most of the continent. The teaching of local values and culture can best be facilitated in a dialogue between teacher and students, which does not occur if an unfamiliar language is used in school.

Sen (1999) argues for a human capability approach to human development and argues that education is essential to the development of children's capability to engage with their local societies and with the challenges of global integration. Being educated has been described by Sen (1999), Carol Spreen and Salim Vally (2006) and Geo-JaJa (2013) as basic capability based on the functions of doing and being. Having access to an education and being knowledgeable allows one to prosper and provides a foundation for other capabilities (Robeyns, 2006). Furthermore, Ingrid Robeyns (ibid.) argues that the capability approach is not a substitute for the human capital approach and human rights approach, but it deepens and broadens the perspective. Seeing education as a right has put the focus on access to education, with free education. Everybody has a right to an education (Spreen & Vally, 2006). However, Sen (1999) argues that various functioning outcomes and achievements exist in a broader normative framework for the evaluation and assessment of the well-being of individuals. The capability approach is in principle multi-dimensional and comprehensive, and therefore accounts for the intrinsic and non-economic roles that education plays (Robeyns, 2006).

The capability approach provides the possibility to evaluate the importance of education, the design of policies, and proposals about the relationship between formal learning and social change, as well as several aspects of people's well-being such as inequality and poverty. Martha Nussbaum (1998) argues that a good and just society should expand people's capabilities. According to Jean Dreze and Sen (2002), "Education is important in the capability approach for both intrinsic and instrumental reasons"; thus the capability approach acknowledges the multiple roles and functions of education in society. Being knowledgeable and having access to an education that allows one to flourish is generally argued to be a valuable capability (Nussbaum, 1998; Unterhalter, 2003) and can be crucial for the expansion of other capabilities.

Geo-JaJa (2013) argues that applying a capability approach to rights in education will expand freedoms of choice and voice in order to stimulate full participation in community decisions. An alternative approach to education aimed at capacity development is described by Geo-JaJa and Steve Azaiki (2010): "In combining different delivery mixes and methodologies, it ensures agency and well-being among the poor and in local institutions; it creates capabilities to improve livelihood functions". This comprehensive approach to education will provide the people with the capacity to develop and contribute to society (Nyerere, 1968; Sen, 2004; Nussbaum, 2011).

To summarize, important theoretical considerations are often ignored in debates about the LoI, including to what extent the valuing of local contexts and cultures is dependent on learning in a local language. The reasons for the increasing use of English in Education for all levels have to do with misplaced associations of development with modernization, where emulation of Western development and

Western educational systems are regarded as the way forward. The theories of Freire and Sen discussed in this chapter incorporate the importance of using a local language and emphasizing the development of local capacity on local terms in Africa and Asia. Such a multiple approach emphasizes the importance of local learning and concepts, articulated in their natural environment, which will give independent nations the power to develop in the new globalized world. This will result in meaningful educational systems for self-reliance, empowerment and self-directed development. The use of local languages and contexts in education will provide to every child the capability to access a quality education and to contribute to development of her or his nation.

3
Development Aid in Education

This chapter reviews development aid in education in Africa and in Asia, with a focus on the historical role of international donors in recipient countries. The picture that emerges from this exercise is counter-intuitive; instead of producing positive outcomes, development strategies have actually worsened the problems. The top-down strategies of the international donor agencies have entrenched mechanisms of exploitation and dependency. According to Steve Klees and Omar Qargha (2013), Roger Riddell observed as late as 2007 that in the case of official development assistance (ODA) programs, although there is evidence for both significant success and noticeable failures, we still do not know if most programs have worked or failed. We have our work cut out for us. Any true assessment of the role of international aid in education in Africa and Asia must begin with the clarification of the fundamental issue of the very purpose of education. Geo-JaJa and Azaiki (2010) point to a lack of common understanding of the relationship between donors and recipients, and to relational impairments that amount to "voicelessness and powerlessness" on the part of local players and to poor education. Thus, what Africa and Asia need is empowering local and global literacy within education. This will enable these regions to break out of the vicious cycle of powerlessness, poverty and marginalized dependent development. Education that holds its own inherent value is at the center of asset building which expands the real freedom that nations value for both social and economic sustainability. For example, literacy is essential for participation in economic, political and social life, but which kind of literacy are we imparting to children?

David Pearson (2007) has reminded us how important literacy is, however literacy alone is not sufficient for quality education; it must be linked to subjects that require logical reasoning and verification, such as science, and to those that promote skepticism and facilitate the exercise of agency. This chapter critically discusses aid modalities and architectures and highlights the principal drivers and maintainers of aid effectiveness in Africa and Asia, building on a holistic approach.

Historical background of development aid

In the immediate post-war years up to about 1950, donors viewed aid as an instrument for supporting developing nations' economies, and aid often aimed at attracting investment and technical expertise that would lead to Western-style industrial development. The international aid system originated in the Marshall Plan, which had the purpose of reconstructing war-ravaged Europe through the agency of the Bretton Woods institutions – the International Bank for Reconstruction and Development (IBRD), popularly known as the World Bank, and the IMF. Although this model was successful in the case of many European countries, the outcomes were less impressive in Africa and Asia when these programs were exported to the latter after independence. The exportation of the model to the non-European developing world emerged out of the decolonization discourse of the 1950s and 1960s and the lack of industrialization in the newly emergent states. The Bretton Woods institutions originally formed for post-war reconstruction in Europe now became instruments of action on development in the emerging countries of Africa and Asia. But the lopsided power relationship between these powerful Western-controlled institutions and recipient countries ensured the prioritization of the needs of donor countries rather than those of the recipient nations. For instance, donor countries sought to achieve their Cold War-era strategic imperatives through attaching well-crafted conditionalities to aid, enforced by the Bretton Woods institutions. This state of affairs crystallized fully during the 1980s when declining commodity prices and concomitant declining revenues forced the recipient countries to seek credit facilities from international creditors to fund their development plans. The conditionalities for these loans came in the form of de-investment in education, privatization and deregulation, among many other stringent requirements imposed by the structural adjustment programs (SAP). Following the general lack of positive results from these programs, world governments in 2002 came together under the auspices of the United Nations in Monterrey, Mexico, to issue the Monterrey Consensus – a recommendation that each of the Organisation for Economic Co-operation and Development (OECD) countries commit 0.7% of its gross national income in aid to developing nations. In 2005, 91 countries made a joint agreement entitled the Paris Declaration of Aid Effectiveness. Rather than focusing on the principles of recipient ownership, alignment, harmonization, managing for results and mutual accountability, the Paris Declaration equated growth with development (OECD, 2008). As currently designed, development aid creates serious disparities between donors and recipients and mitigates against improvements in well-being, standard of living and quality of life across recipient countries. As expressed by Dambisa Mayo (2009, pp. 66, 75), this has led to a culture of "aid dependency" or "addiction" fostered by the international aid complex, which employs half a million people (Klees, 2013, p. 17).

Development no longer seems to be a pressing mission of the OECD, a unique forum in which some governments were at the forefront of efforts to understand and to generally help developing countries heed the new challenges of economic, social and political rights encountered during the period of globalization. Nonetheless, there is growing recognition of the links between giving up the right to self-determination and deprivations of capabilities on the one hand, and (under-)development on the other. In spite of the aid being given, many OECD member countries and multilateral donors now look at the issue of human rights more strategically and the quality of aid partnership more broadly. While some countries such as the Scandinavian countries have adopted the Rights-Based Approach to development aid, others have not yet been inclined to see the benefits of ownership in aid programs or of recipients' human rights linked to jurisdiction over education programs.

The argument here is not that the Marshall Plan should be replicated in Africa or Asia, but many of its principles are relevant. The processes and organizing principles that brought success to the Marshall Plan suggest a much better and more coherent aid approach than is currently available for addressing issues surrounding aid effectiveness. But the literature such as William Easterly (2008), Samoff (2009), Geo-JaJa and Azaiki (2010) and Klees (2013) point out that aid is heavily influenced by the self-interest and political calculations of donors. Much of OECD aid is delinked from the Paris Declaration of Aid Effectiveness principles of 2005, identifying the main sets of rationales for giving aid as: altruism (alleviating human suffering, building and improving the standard of local capacities); strategic geopolitical and security interests (military stations, the "war on terrorism" and the "war on drugs"); political interests (support for international and domestic policies with former colonial territories); and economic interests (export expansion, employment generation, access to resources such as oil and strategic minerals, and tied aid). Some of these motivations for giving aid are not consistent with human rights and with building basic capabilities linked to well-being (see Babaci-Wilhite & Geo-JaJa, 2013).

In fact, shifting motives for giving aid in the past are reflected in the history of aid from members of the OECD Development Assistance Committee (DAC). The member countries of the OECD that allow for comparison of policy experiences with Brazil, Russia, India, China and South Africa (BRICS) work to make domestic and international frameworks seamless. Historically, for some donors, aid serves a multitude of objectives. For others, the quantity and quality of aid are largely shaped by human rights concerns and recipients' development needs. In contrast, larger donors, particularly the United States of America, use aid principally as a stabilizing mechanism for foreign policy (Babaci-Wilhite et al., 2012).

The change in both aid instruments and the underlying ideology of aid donors has two main points: transitional changes and paradigm shifts in the pattern of aid. More recent trends show ideological shifts from state-directed development to a vision of market fundamentalism. In the last decade, balancing aid has not been a problem for newly emerging aid donors like the BRICS countries. Indeed, aid conditionalities led to a loss of ownership of recipients over their own development. The situation is different from aid provided by Nordic countries, mainly Norway, Sweden, Denmark and Finland, in that they are catalysts for human rights, social justice and sustainable development, while OECD aid architecture is determined mainly by political conditionalities; the Nordic countries top the list in the Commitment to Development Index (see Babaci-Wilhite et al.,

2014). The latter pays attention to aid quality and penalizes both aid conditionalities and tied aid. Nordic countries and the BRICS countries consider human rights and the eradication of extreme poverty and hunger to be moral and socio-economic imperatives of donors. But clearly, political and strategic motivations such as security goals, diplomatic ties and access to military bases and strategic minerals have overwhelmingly come to be the price paid by aid recipients. It is therefore important to critically assess whether the next decade will see the traditional aid system just muddying the situation or whether BRICS countries will demand and achieve major reforms in aid. The alternative is that developing nations will continue to be a laboratory for scholars and a battleground for traditional donors (European Union, 2012, pp. 8–9).

The challenge for education aid

The rights-capability approach to education, in contrast to the World Bank's donor policies, includes demassification and improved quality of education. This should be accompanied by a process of inclusion of historically excluded people. In the context of rethinking conditionalities, structural and fiscal stabilization and trade liberalization expanded by donors, the World Bank and the IMF have led to costly education with the lowest rate of economic return. Other serious concerns are associated with quality, equity and effectiveness, as has previously been discussed. This is important for unlocking the full range of individual human capabilities (Alston & Robinson, 2005), and it is an important factor in enlarging indigenous ideologies and democratizing education. The goals of quality and effectiveness, essential contributors to a more holistic approach, are essential. There is little counter-evidence to suggest anything contrary to this assertion, except for evidence based on misleading cost-benefit analysis (see Geo-JaJa, 2004). Therefore, it is about time that recipients shift back to the fundamental values of education in order to empower those whose basic economic, social and cultural rights are violated by the current education architecture.

Education empowers citizens to become leaders and exercise agency. For the marginalized and excluded, it restores their position in the nation-building equation, as it enables them to make choices concerning their desired livelihood. Since the World Conference on EFA in 1990, the realization of the stated aim to bring about "universalized primary education and to massively reduce illiteracy by the end of the decade" has been a daunting task. The Framework for Action to Meet Basic Learning Needs established six goals for the year 2000: Universal access to learning; a focus on equity; emphasis on learning outcomes; broadening the means and the scope of basic education; enhancing the environment for learning; strengthening partnerships by 2000. But the shortcomings of internationally packaged and designed conditionalities have raised discontent and impelled over 110 million children into the labor force, most of them in Africa (UNESCO, 2009). With the World Conference on EFA more than two decades old, more than 300 million children are still likely to have little or no access to education of reasonable quality (UNESCO, 2009). In late January 2014, the cultural agency of the United Nations (UN) issued a report about falling education standards in the world. The report pointed out that a quarter of a billion children worldwide are failing to learn basic reading and math skills in an education crisis that costs governments $129 billion annually. The report made a far more dismal observation about education in sub-Saharan Africa. For that region, the report noted that four in ten African children "cannot read a sentence".

Today's inequalities in education are tomorrow's disparities in competitiveness, capabilities, income and functional integration into society. In some communities where children are deprived of their right to even poor public education, some of the poorest people have to pay for the privilege of sending their children to private schools that lack qualified teachers, books, pencils, clean water and adequate curriculum (see Babaci-Wilhite, 2010).

Despite these challenges and constraints, another role of empowering education is that it opens avenues of communication that would otherwise be closed; it expands personal choice and controls over one's environment; and it is a necessary and sufficient condition for acquiring many other rights. It gives people access to information, equips them to cope better with work and family responsibilities and strengthens their self-confidence to take part in community life. These contributions of schooling were seen as not only unfavorable to the decentralization and privatization of education, but also to be against the policies of donor organizations. This neoliberal framework combines an implausible and ideologically driven faith in markets with a failure to confront the real challenge – building human rights into schooling and rejecting African initiatives that offer true commitment to quality and equity (UNESCO, 2007). This is why the demand for rights in education matters.

The all-encompassing World Bank mandates and imposes donor conditionalities, replacing the public with markets in education that not only splinters societies economically, but also results in widening social disintegration. The market metaphor, riddled with imperfections, inertia and bureaucratic entrepreneurship, introduces budget pressure into human development. Indeed, not only is its outcome in stark contrast to aid that recognizes the values and contribution of right in education, it is unfavorable to the promotion of equitable, sustainable education development that benefits all citizens (Babaci-Wilhite et al., 2012). A number of other roadblocks need to be addressed if education aid is to make any significant contribution to quality development.

Aid effectiveness in education

Human rights in education and in development have not received sufficient attention over recent decades in Africa. The insensitivity of neoliberal reforms, such as the SAP, to the environment and the discarding of local initiatives have had direct negative impact on rights in education (Geo-JaJa, 2014). As discussed above, the initiatives of aid donors ignore the importance of rights in education to human development. This further implies that human capital formation likewise fails to acknowledge that aid is conditioned upon market politics supporting SAP. While education is recognized as a component of anti-poverty and capacity-building programs for sustainable development, education is nevertheless given a low priority on the OECD aid agenda, although it is high on the social, economic and political agendas of aid-receiving nations. At the same time the framework for promoting inclusive education, presumed to be a catalyst for regeneration, has led to capability deprivations (World Bank, 2008; Samoff, 2011). For this reason, human rights must be integrated into aid allocated toward development, and a rights-capability approach should be at the core in defining aid development.

Education aid in Africa and Asia, judged by traditional criteria or by international efficiency indicators, still remains behind that of other regions. For instance, privatization and the imposed economic transformation of the 1990s led to minimal state intervention and left a profound mark on the system of education in Africa, contrary to fostering positive changes and creating new human capabilities. The liberalization of education by imposed structural adjustment conditionalities created a market-oriented education system that became a mechanism of marginalization, as it had little relevance to African and Asian development. The essential truth is that decentralization and marketization components of the new educational discourse – humanistic principles, people-centered pedagogy, and diversification of curriculum and teaching materials – have remained extremely vague (Silova, 2009 quoted by Klees, 2013). Apparently, this neoliberal approach, which has its origin in colonization and which does not incorporate human rights in curriculum, is often in conflict with the need for local ownership, or is caught up in the conflict dynamics of globalization and economic liberalization.

Market-centered reform in privatizing the knowledge sector and the liberalization of schooling intensify disparities in opportunities, exacerbate income polarization, push communities into the underdeveloped Third World and generate socio-economic crisis. Indeed, unless a strong synergy emerges between local communities and structured aid from both bilateral and multilateral institutions, little gain will be made in terms of the functionality of education under/through aid. It was on this basis that Mosha (2006) argued that in Tanzania education aid is not bringing children into quality school systems, since donor aid has basically expanded education in convoluted syllabuses and overloaded programs, with ill-equipped teachers, weak management and poor communication capacity. Just as with the colonial pitfalls of education, all too often schooling today is not an instrument to promote and strengthen learning, as well as to empower citizens and parents to demand greater freedom and critical thinking, but is rather used to reinforce dependency, poverty and discrimination, thus pushing societies toward instability. Geo-JaJa (2010) and Samoff (2011) noted that if there is to be anything like equality of opportunity, it is wrong to justify provision of facilities for some and not for others, or to relegate human rights in education reform to commodification of education. Furthermore, if education is to be universal and compulsory, a concern for equity requires that it should be free, and common sense demands that it should be of sufficient quality for individuals to secure human rights (Babaci-Wilhite & Geo-JaJa, 2013).

Donor aid support is not linked to quality education and therefore does not foster rights in education. It does not give a basis for the empowering of citizens to ensure that their human rights and the human rights of others in their community are respected, protected and fulfilled. Although there has been progress toward this end, the overarching message from the 2011 EFA Global Monitoring Report is that most of the Millennium Development Goals (MDGs) will be missed by a wide margin. Joel Samoff identified one approach to education aid: the construction of aid programs in which "with the foreign funding come ideas and values, advice and directives on how education systems ought to be managed and targeted" (Samoff, 2003, p. 440). External aid has imposed substantial indirect and direct pressure to accept alien policies and educational models. According to Steve Klees (2013), Karen Mundy (2006, p. 18) points out that aid to education exerted direct and indirect control on policy, architecture and programs. Furthermore, advocating for unequal partnership at the expense of control over education is not so very different from the way colonizers organized education (Mason, 2011, p. 446). Just as the colonizers disempowered Africa, recipients of aid are constrained in their ability to undertake human capacity-building.

Donors tend not to allow space for indigenous or localized education. This means that humanistic education rooted in African culture and settings, not currently considered, should be reintroduced into schooling. Education for Africa must achieve a synthesis of its own values first, and then of universal values second, in order to bring local context to the critical mass of knowledge (Crossley, 2012). This will be essential to the effort to promote transformation and achievement of sustainable social and economic rights.

A range of authoritative sources has unambiguously demonstrated that a good deal of OECD aid promotes education commodification and goods dumping (Geo-JaJa & Mangum, 2000). The assessment is that a systematic difference exists between the like-minded countries commonly regarded as committed to international conventions on human rights that bind governments – the Nordic countries – and the OECD countries, whose aid is determined by the market. The former have distinguished their aid architecture from larger OECD donors that use aid as a foreign policy tool or an imperialist penetrating mechanism for dumping surplus products, or for vote getting at the UN (see Andersen, Harr & Tarp, 2006). This point is well articulated by a former Director-General of UNESCO, Amadou-Mahtar M'bow, who noted that development aid should not be reduced to imitation of donor societies, and that development can only come from within. He stressed that development must be endogenous, thought out by people themselves, springing out from the soil on which they live and attuned to their aspirations, the conditions of their environment, the resources at their disposal and the particular genius of their culture These facts underscore a sad message, that Africa did not achieve Universal Primary Education (UPE) by 2015, or reach those children who remain excluded from education due to the complex nature of deprivations and inequities associated with gender, ethnicity and location. The push for economic considerations alone in education becomes ever more difficult, because it results in increased societal segmentation and a new form of imperialism.

It is the quality of aid, and not the issue of quantity, that is essential to quality education for the well-being of people or the promotion of sustainable development. According to Sen, the motive of aid should not simply be to increase economic growth or to accelerate adaptation toward global North-like market economies; the main purpose of aid should be to spread freedom to the unfree. In Sen's thinking, aid of the former kind that concentrates on economic growth alone results in systematic variations between the donor's authoritarianism and the recipient's desired liberation. However, the moral debate on performability of development aid beyond education is captured in Jean-Claude Berthémlemy (2006) who stressed the significant failure of aid to achieve quality education and illustrated that it is an instrument for inculcating a culture of dependency and oppression. High and rising pupil–teacher ratios, overcrowded classrooms, noncontextualization of schooling and lack of female participation are inconsistent with the notion of rights in education.

Aid remains a foreign policy enabler, rather than an enabler of human rights and inclusive education, which are crucial foundations for the resolution of critical social and economic problems. Rather than aid fostering true development or prompting a resurgence of focus on education in localized form, it has only a global focus and results in different forms of human deprivation (Tandon, 2008). There is still no reason to be optimistic, since aid continues to be unrelated to conditions that prevail in Africa and continues to fail to prepare citizens for any adjustment to challenging social and economic forms.

To summarize, Aid to Africa and Asia has been used as an instrument of poverty reduction and economic transformation. However, donor-driven conditionalities through educational aid projects have undermined the ownership of development. Development can only come from within and therefore the real purpose of the Marshall Plan will not be achieved in Africa nor will it be achieved in Asia. Reflection and debate on types of aid support to Africa and Asia suggest that aid support must be seen in the broader context of the integration of social, economic and political rights into education and its future development. This chapter has argued for a new arrangement for aid architecture that is effective and efficient and embodies the dynamics of shifting power from distorted development frameworks. This appears to be the central challenge for African and Asian transformational, sustainable change – a complex whole with roadblocks This has to arise from the cultures, economies and politics in question, and is inconsistent with that of donors.

It is clear that aid practices toward Africa and Asia are predominantly instrumental, inhibiting compliance with stipulated commitments of international conventions and declarations. If Africa and Asia are to develop on their own terms, they will need to rethink jurisdiction over, and conceptualization of, aid. It is imperative that education not be structured through a donor model that has little connection to life in developing countries. The main tasks of donors should be to restore the economic authority of recipients, to provide quality aid, to support human rights policy directions through joint considerations, and to untie aid conditionalities.

The recognition and reflection of the inherent and sovereign rights of nations to exercise self-determination over education and development programs contributes to the efficiency (human capital), effectiveness (capabilities) and sustainability of development. This chapter has critiqued the present state of education aid in which the donor community, under the direction of the World Bank/IMF, formulates and imposes the implementation of structural adjustment-type policy in education. This control constitutes an exercise in domination and exploitation. Aid-receiving nations insist that the present aid practices and development paths mapped out for them by donors, limiting the freedom of citizens and their sovereignty rights, has had disastrous consequences. There is a need for African and Asian countries to take jurisdiction over their education and follow an endogenous development framework that promotes self-determination. In critically assessing rights in education and educational aid that is currently unresponsive to immediate challenges and needs, education aid must break with past mechanisms of aid delivery. Education should be regarded as an inherent right and an important component of self-determination. Development programs must incorporate the goals of enabling human capabilities and emancipating communities from donor political, economic and cultural penetration. Quality education is an essential foundation for nation-building and should be a central focus of education aid.

4
Educational Aid in a Human Rights Perspective

In the previous chapter I explored problems and issues with development aid to Africa and Asia. In this chapter I focus on the application of a human rights perspective to aid in education, making the case that the use of local language as a LoI is crucial to quality learning and self-directed development. A systematic integration of rights into development programs is essential to the improvement of education and the amelioration of poverty's multi-dimensionalities (Babaci-Wilhite et al., 2012). To put it succinctly, there is a need for a paradigm shift in national education policies for African development grounded in the Rights-Based Approach. The Rights-Based Approach can contribute to arresting the processes of exclusion and marginalization that are impacting on the lives of so many children in Africa and Asia in the wake of globalization.

Human rights in educational aid

Unfortunately, international human rights laws, treaties and political commitments have not prevented millions of children from remaining out of school in Africa (UNESCO, 2009). Education aid, which sometimes has no moral ideal or is presumed not to have legal responsibility to aid recipients, is the compounding problem and further does not recognize fully the inherent rights to quality education. Market fundamentalism significantly narrows the valuation of local cultures and contexts. Linking education with marketization and privatization reduces educational opportunities for marginalized children (Babaci-Wilhite & Geo-JaJa, 2013). Marketization denies moral choices in aid and ignores the intrinsic value of education as an end in itself (Sen, 1999, p. 292; Nussbaum, 2000, p. 11). The Asian Tigers history tells us that the successful distribution or provision of education should not be relegated to the market but rather should be controlled and funded by the government (Babaci-Wilhite et al., 2012). As Asia and Africa integrate into the globalized knowledge economy, they should emphasize quality in education rather than expanding quantity, which generates imbalances between the mix of schooling outputs and the needs of the nation. A focus on quality education is critical to the achievement of human rights and should be emphasized in aid. Despite all the claims of human rights in the dialog between donors and recipient countries, repercussions of education aid on public resources dedicated to education prove the contrary. Distributing aid with conditionalities imposes global practices on the local, culminating in enormous human costs, in contrast to unlocking the door to self-determination.

There is a growing consensus that, were human rights principles put in place strategically, they would help to promote and protect people-sensitive approaches, tackle the root causes of poverty and justify a system of schooling that places people at the center of the solution to the challenge to change. An insignificant link exists between aid allocation and the global objective of supporting priority human development goals and capacity-building (see Oxfam, 2010). The purpose here is not to indict donors but to put forward the need to integrate human rights into education aid for sustainable development (OECD, 2010). This point has been obscured by focusing more on income poverty and income inequality, to the neglect of deprivations that relate to other variables such as respect for human rights, social inclusion and the promotion of a range of cultural objectives.

Rights in education that supports enlarging opportunities, enhancing participatory freedom and reasonable expectations of the citizen's well-being should be central to an effective aid policy (Sen, 1999). Furthermore, Sen (1999) contended that valuable activities that emerge from rights entitlements in education ought to be a good enough reason for donor compliance with the integration of human rights into development aid. In more specific terms, rights in education that includes elements of availability, accessibility, acceptability and adaptability (the 4As) should be a common denominator of education in all its forms and at all levels (Tomasevski, 2006). However, the distinction made between rights to education and rights in education is important for elevating education aid to the moral high ground. Many of the criticisms of education approaches apply more readily to the narrow concept of market fundamentalism than to the more encapsulating capability deprivation in explaining indicators of distributive inequities. In a broader perspective, rights in education take account of the economic, cultural, political and social background of locality in identifying and curricularizing students' needs and concerns. This critical, localized and flexible education curriculum incorporates local needs, protects human rights and promotes social and cultural activities of various kinds. With human rights incorporated into aid, education strategies become a source for combating capability deprivation and economic inequality. In this new framework, learning produces a virtuous circle (for more detail, see Babaci-Wilhite & Geo-JaJa, 2011). With this understanding, a Rights-Capability-Based Approach becomes imperative for the making of education policy. This justifies an earlier remark in this chapter about the obligations of aid donors to relinquish control over education on the principle of participation, inclusion and consultation rooted in human rights so as to promote and protect culturally sensitive approaches.

A paradigm shift in local language and educational aid

A paradigm shift must happen; otherwise schooling can be deadly, in the words of Katarina Tomasevski's (2004) work on rights in education. Furthermore, she argues that implementing a Rights-Based Approach in education requires the identification and abolition of contrary practices (Tomasevski, 2004, paragraph 50). This is rendered difficult by two assumptions: one, that getting children into schools is the end rather than a means of education; and two, an even more dangerous assumption that any schooling is good for children. In Tomasevski's views (2004, paragraph 5), the impact of Rights-Based education should be assessed by the contribution it makes to the enjoyment of all human rights. Accordingly she argues that "International human rights law demands substitution of the previous requirement upon children to adapt themselves to whatever education was available by adapting education to the best interests of each child" (Tomasevski, 2006). The right to use one's own language is denied if children lose access to it during the educational process. The securing of this right should be seen as a first step in developing a global monitoring system for education programs for all. Tomasevski (2006) argues that these linkages have been betrayed not only by donors, but even by the UN. Moreover, while the right to education – like all human rights – is universal and inalienable, several conventions have led to its enshrinement in various constitutions as a right of the people, thereby placing binding responsibility on ratifying states. The education of children has to fulfill further demands that can be made on quality education; these include, in the words of Tomasevski, that "a rights-based analysis of poverty is crucial to identify where poverty results from denial and violations of human rights" (2006, paragraph 17).

An engagement with human rights makes collective action feasible and productive, seeking to create a more structural and holistic approach to quality educational development for all. With this new approach, communities, organizations and governments have an opportunity to rethink and pursue innovative education but also the 4As. This is not simply a function of having enough schools, textbooks, teachers and infrastructural facilities; it is a result of the right to social and local context in schooling in which all people have the freedom to improve their economic, political and social welfare and participate in their respective desired cultures of well-being.

Rights in education refer to the ability of a nation to sustain its education without outside intervention. The importance of redefining education rights cannot be overemphasized. In assigning high priority to human rights in both aid allocation and the choice of aid modalities, we need to relate educational rights and the economics of poverty to intangible resources such as language and cognitive capabilities. This clarifies the role of dominant languages and indigenous language in formal education in a context of social mobility at a global level. Sen's conceptualization of poverty as "capability deprivation" is highly relevant to the LoI debate. The central question in reducing poverty is to identify the most critical and cost-effective input to change the conditions of poverty, and to address these through expanding human capabilities. Part of the explanation for teaching in a dominant foreign LoI is a parental belief in many parts of Africa and Asia that education in English, for example, will increase work prospects and social mobility for their children and thus provide a way out of poverty. However, for the large majority of the children, English LoI does not lead to the promised outcomes, as countless research results show, for example those of Brock-Utne (2000), Kwesi Kwaa-Prah (2005) and Qorro (2009). Indigenous language as a LoI would in most cases be a better way to achieve real capability-building, in Sen's sense. Moreover, based on study of the economics of poverty, not using the indigenous or minority MT as the main medium of education perpetuates poverty.

The UN agencies with oversight of the MDGs, UNESCO and UNICEF, should undertake an effort to inform and educate states about:

- Firstly, the negative consequences that result from forced instruction of children in foreign language-medium education.
- Secondly, that education in indigenous language has significant positive outcomes in addressing poverty, building capacity, strengthening education, and eliminating inequality and social marginalization.
- Thirdly, failure to apply the relevant human rights standards in this area can result in serious violations of state obligations in Tomasevski's (2004) words.

Concerning the place of language rights in the education aid of OECD member countries and multilateral donors, I turn to the issue of human rights more strategically and the quality of aid partnership more broadly. While some countries, such as Norway, have adopted the Rights-Based Approach to development aid, others have not yet accommodated the importance of aid program ownership or of recipients' right of jurisdiction over education programs (Babaci-Wilhite & Geo-JaJa, 2013).

Despite the Universal Declarations of Human Rights (UDHR), education rights are still works in progress in developing countries. Moreover, while the right to education – like all human rights – is universal and inalienable, several conventions have led to its enshrinement in various constitutions as a right, thereby placing binding responsibility on the ratifying states. This complexity serves as a serious threat to a common understanding of the unbalanced relationship between donors and recipients, and the challenges of integrating indigenous knowledge and indigenous languages as human rights in education. Education should be a human right in itself and an indispensable means of realizing other human rights.

Donor aid today is not linked to quality education. Higher-quality education requires not only education about human rights but education for human rights – citizens should be empowered to take action to ensure that their human rights and those of others in their community are respected, protected and fulfilled. Although there has been progress toward this end, the overarching message from the 2011 Education for All Global Monitoring Report is that most of the MDGs will be missed by a wide margin. MDGs are grounded in the recognition inherent in rights to quality education, that basic education is a human right and a vital part of an individual's capacity to attain greater freedom – in other words, for people to lead lives they value. Indeed, the exaggerated emphasis placed on the opportunity costs of schooling does not account for cost recovery, user fees, transportation costs and the benefits of education that are conditioned on context (Samoff, 2009; Geo-JaJa, 2013).

According to Arnove (2012), education must achieve a synthesis of own value first, and of universal values second, in order to create context in the critical mass of knowledge and attitude necessary for obtaining transformation and recovery of sustainable social and economic rights. Furthermore, Geo-JaJa argues that a range of authoritative sources has demonstrated that a good deal of OECD aid promotes education commodification and goods dumping. As Geo-JaJa and I explored in a recent paper published in 2014, a systematic difference exists between the Nordic countries on the one hand, which are commonly regarded as committed to international conventions on human rights that bind governments, and the other OECD countries on the other hand, which use aid as a foreign policy tool or an imperialist penetrating mechanism for dumping surplus products, or for vote getting at the UN (see also Geo-JaJa & Azaiki, 2010).

Carnoy argued that marketization and privatization of education, which commodifies knowledge and learning activities, leaves children with educational opportunities of poor quality. Furthermore, Samoff claims that, as Africa integrates into the globalized knowledge economy, the emphasis should be on quality in education, rather than quantitative expansion, which generates imbalances between the mix of schooling outputs and the needs of the nation. The conceptualizing of learning has implications for ways in which human rights in aid are interrelated to rights in schools rather than right to schools.

In many parts of Africa and Asia, the initial gains made following "decades of development" have disappeared, resulting in disintegration and dependency. Hence, the mainstreaming of human rights into education aid and strengthening the ability to convert them into valued functioning is imperative. Repackaging aid to fit or go beyond the Paris Declaration of Aid Effectiveness of 2005 is to strengthen aid for greater support for rights in education. Samoff (2007) asserts that the high and rising pupil–teacher ratios, overcrowded classrooms, non-contextualization of schooling and female absence from classroom participation are inconsistent with the notion of rights in education. Furthermore, drawing on the work of Nussbaum (2000), aid interventions require focus on education and cannot be alien to or separated from rights in education, because the latter will propagate the extension of quality education to the marginalized. Such a broader perspective challenges the neoliberal focus on instrumental processes that predominates to the detriment of localized knowledge and learning that are fundamental to adjustments and resolution of everyday problems. In line with Nussbaum, Inga Bostad argues that a focus should be placed on the role of education as an instrument of empowerment and voice for community and thus an important contributor to securing human rights. The logic and supportive evidence for the rights discourse abound in the UDHR, 1948; the European Social Charter, 1961; the African Charter on Human and Peoples' Rights, 1981; the Convention on the Rights of the Child, 1989; and the 2005 World Summit. These conventions give assurances that the implementation of human rights in aid make for a more holistic framework for promoting rights in education for self-determination and sustainable development. The roles of suppliers and aid recipients are made explicit in these international treaties, which maintain that quality education shall be compulsory and free to all. Quality education, a key mandate in the above universal principles, is absolutely essential for a positive relationship between rights and aid for sustainable development. Scaled-up aid has denied or underestimated the intrinsic values of local knowledge and universal principles of human rights.

Human rights in aid for sustainable transformation and development

The subject of human rights is both an objective in its own right and a factor for sustainable development and sustainable social transformation. It should not be alien to development aid. Human rights should be included in any analysis of the social, economic and political foundations for change toward emancipation from deprivations. But traditional aid not only delinks human rights and intrinsic norms from practical aid missions, it also assigns low priority to human rights (individual) and sovereign rights (nations) (Babaci-Wilhite & Geo-JaJa, 2013). This is seen as constitutive of the multi-dimensional definition of poverty and variation in living conditions within and among countries.

It is evidenced by Sen (1999) and other noted experts in the field, including Nussbaum (2000), Robeyns (2006) and Tomasevski (2006), that quality education, a key mandate in the above universal principles, is *absolutely* essential for a positive relationship between rights and aid for sustainable development. Grudgingly, the World Bank focuses on efficiency to the exclusion of other considerations. It states that aid conditionality or aid modalities can result in severely mismatched development, as aid is often designed to serve the strategic and economic interests of the donor (see World Bank, 2003). Aid-giving has driven most well-resourced school-aged children away from dysfunctional public schools and into private schools in many countries in Africa and Asia.

To summarize this chapter, a paradigm shift in the present practices of educating children through the medium of foreign or state languages is crucial, as these practices are completely contrary to solid theories for quality education as demonstrated by experts in the field. In addition, they also violate the parents' right to intergenerational transmission of their values and culture, including their languages. Education is not only a right in itself but is also indispensable for the exercise of other human rights. Rights in education, rooted in cultural reaffirmation, are imperative for the systemic involvement of stakeholders to exercise their economic, social and political rights.

Aid programs must address the problem that children in developing nations are being educated primarily in dominant LoI, which has had extremely negative consequences for the achievement of the right to education and rights in education. African and Asian countries should take control over their education and follow an endogenous development framework that aims for self-determination. Education should be recognized as an inherent right and a right of self-determination, and should be given priority in development.

5

A Rights-Based Approach in North–South Academic Collaboration within the Context of Development Aid

This chapter describes and assesses initiatives by Norad to improve and expand collaboration with universities in developing nations through the sponsorship of university partnerships between Norwegian universities and universities in the South. This chapter uses as a case study of collaboration between the University of Oslo and the University of Dar-es-Salaam in Tanzania from 2009 entitled "Programme for Institutional Transformation, Research and Outreach" (PITRO). The overall goal of the PITRO collaboration was to increase the contribution of the University of Dar-es-Salaam to Tanzania's efforts to stimulate economic growth, reduce poverty and improve social well-being through the transformation of the education and science sectors. The core objective of PITRO was to ensure equitable access to high-quality tertiary education in order to acquire necessary knowledge and skills both for increasing economic productivity and for reducing poverty.

The University of Oslo has a long history of North–South collaboration with a wide range of institutions on the African and Asian continents. The chapter situates the collaboration in relation to Norwegian aid, and compares the program with previous collaborations, giving attention to the ways they were developed, how problems and challenges were tackled and the probable consequences for development. The chapter also compares and contrasts the structure of, and experiences from, these programs and points out strengths and weaknesses. Attention will be given to the Rights-Based

Approach, an important new dimension in North–South collaboration on higher education in Africa and Asia. It argues that the incorporation of this approach provides a basis for developing new policies and programs that strengthen African collaboration within research, education and capacity-building in higher education.

I begin with a review of documents, reports and critical discussions of major topics related to the development of higher education in Africa within the Rights-Based Approach. I also draw on my experiences working with the leaders of the projects supported by PITRO in order to shed light on their positive and negatives experiences in relation to the program's objectives. I have also drawn on my own experiences as the administrative coordinator of the PITRO program at the University of Oslo. I traveled twice each year to Tanzania from 2009 to 2012 and developed an understanding of the context for the collaboration and the effectiveness of each of the projects supported by the program. This included attendance at the bi-annual meetings of the University of Oslo and the University of Dar-es-Salaam leadership, conducted in both Tanzania and in Norway. The observation of the decision-making and negotiations on the program design have been crucial for my analysis of the strengths, weaknesses and issues for further development of North–South university collaboration.

The chapter concludes with an assessment of the benefits and problems of this type of development partnership, as well as with reflections on the challenges in doing North–South university collaborations within the Rights-Based Approach. It provides new insights on North–South collaboration aimed at advancing the role of higher education in resolving African development challenges. I hope this provides a basis for developing new policies and programs to strengthen the sustainability of higher education institutions and development models in Africa as well as in Asia.

Norwegian aid in education for development

The collaboration between the University of Oslo and the University of Dar-es-Salaam was supported by Norad, which follows the Paris Declaration of Aid Effectiveness in demanding a people-oriented aid framework based on key principles such as ownership, alignment and mutual accountability. Contrary to the common understanding of aid driven by self-interest or imperialist hegemonic penetration, Norad aid to Tanzania is shaped by mutual benefit and focused on protecting and promoting human rights (Babaci-Wilhite et al., 2013). In this vein, Norway's aid has built capacities and accountability within the different layers of governments (Norad, 2012).

This approach is informed by the belief that decision-making and local jurisdiction in the use of aid resources presents opportunities for addressing human rights and changing the international aid context. The Rights-Based Approach to aid is in line with capacity building and the increase in the numbers of people who complete higher education. However, more recently political conditionality has obtained broader legitimacy; human rights and good governance are now more explicitly tied to aid distribution (Selbervik, 2006a, 2006b). Norway's aid-integrating human rights principles, deeply rooted in equality and mutual rights in development, link appropriately to Tanzania's development and education needs (Babaci-Wilhite et al., 2013). Norad policy has tried to develop programs in line with human rights principles. In light of this perspective, Mason argues that "social justice, the benefit of humankind and the human rights and dignity protection give reasons to the justification of directions of knowledge flow (North–South, South–North)" (2011, p. 3). Mark Mason (2011) has cited some principles as guidelines for research in partnership with developing countries, created in conjunction with his work for the Swiss commission for research partnership with developing countries:

– Decide on the objectives together,
– Build up mutual trust,
– Share information, develop networks,
– Share responsibilities,
– Create transparency,
– Monitor and evaluate the collaboration,
– Disseminate the results,
– Apply the results,
– Share profits equitably, and
– Increase research capacity. These principles have been put in place by Norway in its partnership with Tanzania in order to invite Tanzanian input on the partnership.

The approach outlined by Mason is in line with the Rights-Based Approach, which had its origins in 1993. When the UN held the World Conference on Human Rights in Vienna, the Vienna Declaration and Programme of Action were conceived, linking democracy, human rights, sustainability and development (see UN, 2011). In 1997, the UN Secretary-General called for a mainstreaming of human rights into all work of the UN, and in 2003 various organizations and agencies met to develop the government's responsibility in order to insure the Rights-Based Approach (see Babaci-Wilhite, 2012). The Rights-Based Approach works to shift the development paradigm away from aid and toward moral duty imposed on the world through the international consensus of human rights. Human rights rhetoric was used to develop a "mode of operation" to describe a person's manner of working, their method of operating or functioning, as a driving force to bring effectiveness to human rights (Nelson & Dorsey, 2003). Development cooperation should be regarded as another tool to increase human rights effectiveness as it increases human capabilities, functions and opportunities in societies. Joint programs have the potential to empower if trust and an equal relationship through common understanding is embedded in the programs. This enhances confidence in the collaboration not only at the academic level but at the institutional one. This further leads to the linkage between human rights and development and enables policy makers and developers to incorporate the Rights-Based Approach within the "common understanding", assuring the principles of indivisibility, equality, participation and inclusion (UNDP, 2006, pp. 17–18). Rights are defined as entitlements that belong to all human beings regardless of race, ethnicity or socio-economic class (Nussbaum, 1998, p. 273). All humans therefore are rights holders, and it is the governments' duty to provide these rights. The next step within the Rights-Based Approach is to educate both the rights holders and the duty bearers by articulating the rights of citizens and the duties of the government.

Historical collaboration between Norwegian and African Universities

The University of Oslo began institutional collaboration with African and Asian universities in the 1980s, and since 1991 the Norwegian Programme for Development, Research and Education (NUFU) that

preceded PITRO has provided an important framework for cooperation with institutions in developing countries. The goal of NUFU is to
support the development of sustainable capacity and to build competence in research and research-based higher education in developing
countries. It is expected that this enhanced academic collaboration in
the South–South and between South and North will be relevant for
national development and poverty reduction. The program covered
four five-year NUFU project periods, the last being 2007–2012.

Projects have been developed in several countries in Africa, mainly
Botswana, Ethiopia, Mali, Mozambique, Namibia, South Africa,
Zimbabwe and Uganda. NUFU topics include the environment, education and health. Some projects been financed for two terms, over
the period 2007–2012. A NUFU project has a time frame of up to
two five-year periods, and this has allowed for building up and
strengthening units at the partner universities. Because the units are
built on local needs and development of the institutions' own staff
and academic competency, most of them are sustainable beyond
project completion. The most important partner institutions in the
NUFU program in the project period 2007–2012 were the University of Malawi, the University of Dar-es-Salaam and Addis Ababa
University. Most of the University of Oslo's NUFU projects in this
period were network projects, with more than one partner outside
Norway.

The NUFU program has provided supplementary support to
certain activities connected to projects in 2007–2012, including
the Norwegian University Cooperation Programme for Capacity
Development in Sudan (NUCOOP). NUCOOP supports cooperative projects between higher education institutions in Sudan and
Norway that focus on capacity-building, research and infrastructure.
The NUCOOP program aims to contribute to the development of
sustainable capacity of higher education institutions in Sudan and
South Sudan and to provide the workforce with adequate qualifications. In more specific terms, as of 2009 NUFU projects had reported
55 affiliate PhD candidates and 42 master's degree students.

The Norwegian master's program

Like NUFU, Norad's Programme for Master Studies (NOMA) is also
financed by Norad and is managed by the Norwegian Centre for

International Collaboration in Higher Education (SIU). NOMA is a program that provides financial support to develop and run master's degree programs at universities in the South in collaboration with universities in Norway. The NOMA program is based on equal participatory partnerships between higher education institutions in Norway and in the South. The goal of NOMA is to contribute to the tertiary training of staffs in public and private sectors as well as civil society in selected developing countries. This is achieved through building capacity at the master's level in higher education institutions in the South.

The needs and priorities of the partners in the South are the basis for the collaboration. The University of Oslo has been involved in several NOMA projects: the Southern African master's program in mathematical modeling with the University of Dar-es-Salaam as a main partner, and the master's program in "Nutrition, Human Rights and Governance" with Makerere University in Uganda as a main partner; two integrated programs at the University of Malawi on health and information systems; and an integrated master's in health informatics in Tanzania and Ethiopia with the University of Dar-es-Salaam are some of the main partnerships.

A new model for institutional and transformation research outreach

For several decades Tanzania has been one of the most important countries for Norwegian development collaboration and has been prioritized as a collaboration partner for Norwegian aid. In the past, the focus was on technical assistance and sending Norwegian experts to Tanzania. More recently most of the development aid has been targeted at the social sector, particularly health, education and infrastructure (see Norad, 2012). Other sectors of the country receiving development assistance are civilian organizations such as Norwegian Church Aid, Norwegian People's Aid and Care International, and the UN international higher education development in Tanzania. A collaboration agreement was signed between Tanzania and Norway in 2006 with four main areas of focus: (1) budget support, good governance and human rights, education and culture, and tax for development; (2) climate and environment, including green revolution; (3) energy and oil for development; and (4) health, including maternal health and infant mortality (Norad, 2012). The agreement emphasized that Norwegian aid shall be made visible in Tanzania's Joint Assistance Strategy, which Norway signed in December 2006 together with 18 other donors. The agreement also stipulated that Tanzanian authorities should spend less time on administrating donor aid, so as to be able to spend more time on managing the development of the country. At the same time, the agreement sets clear conditions for Tanzania, including zero tolerance on corruption. In 2011, bilateral assistance to Tanzania amounted to 640 million Norwegian kroner, of which 33 million kroner was dedicated to education.

In line with the Norad agreement, the Norwegian Ministry of Foreign Affairs has revised the modalities for Norwegian support to Tanzanian universities. The administration of support to the University of Dar-es-Salaam was implemented through the agreement between Norad and SIU in 2009. The agreement for the PITRO program with the Norwegian partner institution covered a period of three years, from 2009 to 2012. The PITRO program had an annual budget of 15 million Norwegian kroner per year. The new element in the collaboration between Norway and the University of Dar-es-Salaam was the introduction of a Norwegian partner institution. The objective of the institutional collaboration with the University of Dar-es-Salaam was to facilitate the realization of PITRO program goals. It was stated in the call for tenders that the Norwegian partner institution's participation in the PITRO program would be compensated with a maximum of 10.5 million Norwegian kroner for the period 2009 to 2012 and subject to the contractual obligations of the Tripartite Agreement between SIU, the University of Oslo and the University of Dar-es-Salaam.

The University of Oslo and the University of Dar-es-Salaam have collaborated in research and teaching in a variety of different fields for many years, but PITRO was a more comprehensive program. PITRO was a joint program with the overall goal to increase the contribution of the University of Dar-es-Salaam to Tanzania's efforts to achieve economic growth, reduce poverty and improve social well-being of the country through transformation of the education, science and technology sectors. The objective was to ensure that qualifying candidates have equitable access to high-quality education that should contribute to enhancing knowledge and skills for increasing productivity and reducing poverty. The program involved collaboration between the central administration and a number of different departments at the two institutions. PITRO involves collaboration in three thematic areas: "Education", "Environment and Conservation" and "Good Governance", which are themes under Norad's strategic development goals. It also involves cross-cutting institutional interventions at the University of Dar-es-Salaam and the University of Oslo, which will provide a direct benefit through the outreach dimensions. This joint partnership is in accordance with the University of Oslo's Strategy 2020, which states that the University of Oslo shall be more selective and purposeful in its institutional collaboration and will give priority to long-term cooperation with some of the best international research and educational institutions globally. Collaboration with seven prioritized countries in other parts of the world, including selected institutions in the global South, will be improved. The Assessment Committee evaluated several proposals and approved six research projects – four in environment and conservation, one in good governance and one in the education sector. The budget for each project was 800,000 Norwegian kroner for the University of Oslo and a maximum of 160,000 US dollars for the University of Dar-es-Salaam.

The first year of the program was dedicated to establishing contacts and developing collaborative projects in research, capacity-building and institution-building projects. Proposals for research and teaching were prepared and submitted in a competitive process which led, in summer 2010, to the selection of the six projects previously mentioned to be funded for the next two years. In addition to these projects, the universities collaborate on a range of activities including quality assurance, the library, research ethics and HIV/AIDS. An important priority for PITRO was to align projects with the strategic priorities of the two universities. The University of Oslo recently underwent a strategic process of setting academic priorities for research and education. The University of Oslo identified seven inter-faculty research areas in fields where the knowledge demands in society are complex and require interdisciplinary collaboration. Several of these research areas – the education sector, the environment sector and good governance – were relevant for North–South cooperation generally and for PITRO particularly. The University of Oslo has pursued the PITRO collaboration with great interest, hoping and indeed believing that the PITRO partnership program will be mutually enriching for the University of Dar-es-Salaam and the University of Oslo.

The first component – research, development and outreach – focuses on demand-driven and basic research in order to empower and provide solutions to societal needs or nation-building challenges. However, it was a challenge to find partners at the University of Oslo, therefore only a few projects were proposed and selected.

The *education* sector was represented by only one project, because the University of Oslo did not succeed in meeting the expectations of the University of Dar-es-Salaam for educational projects. The only proposed project that was selected focused on learners with disabilities, and includes research as well as capacity-building activities. The project has two components: one related to training teachers for inclusion of learners with disabilities in primary schools, and the other related to learning strategies among the deaf.

The *environment* sector was the most popular, as there are a great number of experts and researchers in the field of environmental studies at the University of Oslo. One project assessed the impact of climate change and variability on the natural environment and on socio-economic aspects as reflected in rural people's livelihoods in Tanzania. This was one of the most successful projects since it built on a previous collaboration. Another project focused on biodiversity in a conservation program for the Pare Mountains. It is a newly established collaborative project involving two University of Dar-es-Salaam departments and the Natural History Museum, which is a part of the University of Oslo. The project team from the University of Oslo identified the main biodiversity distribution patterns in the least-known part of the Eastern Arc Mountains in order to estimate the implications for conservation and provide recommendations to nature managers and local policy makers. Another project is examining the contribution to the geological and hydrological investigations of past and present groundwater distribution in the Bahi-Manyoni area in central Tanzania. This project was also built on previous collaboration and has been successful in improving understanding of the formation of Kilimantinde beds and associated uranium enrichments, as well as improving water quality. Another project focused on sustainable energy systems. It explored the socio-economic and technical challenges involved in implementing decentralized solar-based mini-grids in Tanzania. Despite delays due to problems in equipment delivery, the project succeeded in setting up solar mini-grids in a village south of Dar-es-Salaam and produced preliminary findings on technical adaptation to local conditions and know-how transfer.

In *good governance*, the only project selected involves international environmental and climate law, and its application in Tanzania through the REDD+ strategy (the UN collaboration program on Reducing Emissions from Deforestation and Forest Degradation in Developing Countries). The project aims at legal capacity-building in Tanzania, through teaching and research, which are related to protection of biodiversity and climate change. A teaching component covered international environmental and climate law and issues of implementation in Tanzania. The research focused in particular on legal aspects of forest protection and the implementation of REDD+ strategy in the country. The research was carried out by Tanzanian master's and PhD students with the support of researchers from Oslo. The project also involves student exchange between the two universities.

The second component of PITRO, institutional transformation and capacity-building, aims at improving capacity of the University of Dar-es-Salaam in terms of human resources, physical infrastructure, strategic interventions, institutional transformation, improved employability of graduates and enhanced performance of staff. The component focused training at the PhD and master's degree levels. Administrative and technical staff participated in short courses locally and internationally to enhance their performance in core mission and administrative functions of the University of Dar-es-Salaam. Part of the improvement of the physical infrastructure at the University of Dar-es-Salaam included the establishment of a gymnasium and fitness center. The cooperation also focused on digitalization of the University of Dar-es-Salaam library, which has been a successful part of our institutional project. Another cooperative project requested by the University of Dar-es-Salaam was the transformation of the University of Dar-es-Salaam gender center from an advocacy center to an academic center, but the project was not fully executed due to lack of time and funds. In the domain of HIV/AIDS, it was decided that the University of Dar-es-Salaam has enough expertise and therefore the University of Oslo would only support academic writing through writing workshops. The examination of the *Quality Assurance* proceedures has revealed that the University of Dar-es-Salaam has greater expertise in this domain than the University of Oslo. Therefore this institutional intervention stopped in an early phase. A net-based course in research ethics has offered fruitful seminars with good results. It will continue through a different cooperation mechanism due to insufficient funding of PITRO.

PITRO, like NUFU and NOMA, is financed by Norad and managed by SIU. SIU has assisted the Royal Norwegian Embassy in Dar-es-Salaam since 2006 with administration of Norwegian support to three Tanzanian universities. The arrangement is known as the Tanzania Agreement. PITRO has been assessed as having contributed to capacity-building at the University of Dar-es-Salaam, as well as ensuring long-term equitable collaboration and research capabilities for sustainable self-development. A recent independent review gave a positive evaluation of the collaboration. The program ended in early 2013, but was restarted under a different name in the autumn of 2013. NOMA and NUFU will be replaced by the Norwegian Programme for Capacity Building in Higher Education and Research for Development (NORHED).

The Norwegian higher education program

NORHED is an educational capacity building program with the goal to build higher education and research skills in mainly low- and middle-income countries through partnerships with Norwegian higher education institutions. This is seen as a means to enhance sustainable conditions conducive to societal development and poverty reduction within the Rights-Based Approach by giving more voices to the South, for example by letting Southern countries decide what they want to focus on and the aims and content of the collaboration. Underscoring the aim of Southern partner ownership and capacity-building and the NORHED principle of equal partnerships, the South-based model is preferred wherever possible. The eligibility requirements for prospective partners is that partners from low- and middle-income countries must be higher education institutions that are accredited/recognized by in-country national authorities in countries registered as OECD-DAC official development assistance recipients, or as listed in the specific call for applications. The Norwegian partners must be higher education institutions accredited by Nokut, the Norwegian Agency for Quality Education, offer accredited degree programs, and operate in accordance with

Guidelines for Quality Provision in Cross-Border Higher Education (UNESCO/OECD, 2005). Other academic institutions or institutes can apply for NORHED projects in partnership with a Norwegian higher education institution accredited by Nokut.

The program goal is sustainable economic, social and environmental development in low- and middle-income countries. The program objective is to strengthen capacity in higher education institutions in low- and middle-income countries and contribute to:

(1) a larger and better-qualified workforce;
(2) increased knowledge production;
(3) evidence-based policy and decision-making; and
(4) enhanced gender equality in the participating countries.

Increased capacity means strengthened capacity of institutions in low- and middle-income countries to educate more and better-qualified candidates, and increased quality and quantity of research conducted by the countries' own researchers.

The outcomes in the NORHED program are to be measured within six identified areas. Strengthening of higher education institutions refers to:

(1) Producing more and better research relevant to the identified areas/sub-programs;
(2) Producing more and better-qualified graduates relevant to the identified areas/sub-programs.

Both components shall be included in the projects. NORHED takes a holistic approach to capacity-building and strengthening of higher education institutions by supporting a range of output-producing activities intended to be combined in the best way possible to produce sustainable results in the long run. NORHED outputs and activities are to be organized under six main categories:

(1) Programs: Increase and strengthen education and research programs;
(2) Systems: Strengthen education and research systems;
(3) Infrastructure: Improve institutional infrastructure for education and research (including supplies and equipment, but not buildings);

(4) People: Increase capacity and competence of staff and students;
(5) Gender: Improve gender balance and gender focus in all education and research programs;
(6) Methods: Enhance methods for effective and high-quality teaching and research.

NORHED has six sub-programs reflecting key priority areas for Norway where capacity-building for education and research in low- and middle-income countries is needed.

These are:

(1) Education and Training;
(2) Health;
(3) Natural Resource Management, Climate Change and Environment;
(4) Democratic and Economic Governance;
(5) Humanities, Culture, Media and Communication; and
(6) Capacity Development in South Sudan.

In addition to the thematic and geographic foci of the sub-programs, NORHED will incorporate additional cross-cutting purposes. One of the principal aims is that gender be a cross-cutting priority in all sub-programs. A gender mainstreaming approach implies integration of gender perspectives in the planning and implementation of all aspects of the project cycle. This includes elements such as design of curricula and research projects, human resources and recruitment, teaching, supervision, research activities as well as monitoring and evaluation. Educational programs and research activities which explicitly address issues related to gender equality are encouraged. Measures should be taken to increase the number of female students at all levels, as well as female researchers, project participants and project coordinators. All projects should make every effort to recruit at least 50% female students at all levels.

NORHED will give priority to countries identified for long-term bilateral collaboration with Norway. These are: Burundi, Ethiopia, Liberia, Madagascar, Malawi, Mali, Mozambique, Sudan, South Sudan, Tanzania, Uganda and Zambia in Africa; Afghanistan, Nepal, Pakistan, Sri Lanka and Timor-Leste in Asia; and the Palestinian Territories in the Middle East. The geographic foci per sub-program are:

(1) Education and training: Primarily low- and middle-income countries in Sub-Saharan Africa, though low- and middle-income countries in the Middle East, Asia and Latin American regions are also eligible.
(2) Health: Primarily low- and middle-income countries in Sub-Saharan Africa.
(3) Natural resource management, climate change and environment: Sub-Saharan Africa, South Asia, Southeast Asia and Latin America.
(4) Democratic and economic governance: Primarily low- and middle-income countries in Sub-Saharan Africa, fragile states and post-conflict states.
(5) Humanities, culture, media and communication: Low- and middle-income countries, fragile and post-conflict states.
(6) Capacity development in South Sudan.

The Rights-Based Approach in Norwegian aid is crucial in order that the collaboration be built on shared values to establish broader and more inclusive partnerships in the acceleration of poverty reduction. NORHED has acknowledged that enhanced capacities should result from programs that respond to the demand and needs of developing countries. These ideas are articulated in both the Norway and Nairobi outcome documents and in the Paris Declaration and Accra Agenda for Action. In the Nairobi outcome document it is stated, "There is the need to enhance local capacity in developing countries... in contribution to national development priorities, at the request of developing countries" (OECD, 2011, p. 3). The Accra Agenda for Action states, "Without robust capacity – strong institutions, systems, and local expertise – developing countries cannot fully own and manage their development processes... Donors' support for capacity development will be demand-driven and designed to support country ownership" (OECD, 2011, p. 3).

Future alternatives and challenges

The focus of Norway's aid partnership research projects and programs should include the Rights-Based Approach. NORHED was established on the principle that its partners in Africa had equitable input to programs and goals. According to Mason (2011), an effective and balanced partnership has been offered as one response to the failures of international development assistance. Furthermore, Mason argues that the mission of aid is to assist receiving countries in their efforts to eliminate poverty and enlarge their capabilities for sustainable development. The global policies of development aid are guided by the MDGs, but they are incomplete and do not account for individual rights and community sovereignty and needs (see Babaci-Wilhite et al., 2012a). Aid recipients that know best their needs and problems are not accorded the opportunity to determine their own development path, to promote local ownership and to support long-term self-determination. Rather, aid is a mechanism for promoting donor self-interests and perpetuating aid dependency, and it has a history of capabilities deprivation in developing countries (Samoff, 2003; Tandon, 2008; Babaci-Wilhite & Geo-JaJa, 2011). These points are taken from the current discourse on the Rights-Based and Capability Approach to development (see Sen, 1999; Babaci-Wilhite et al., 2012) and also from the numerous UN rights declarations. There are two main, separate areas, in which aid is catalytic. First, in promoting growth, aid may lead to changes in policies, infrastructure and institutions. Second, in being complementary to other development inputs, aid should specifically guide international support to localized development processes and problems and be used strategically to reach the poor and leverage transformative change in recipient countries (Geo-JaJa & Zajda, 2005). Aid with conditionality requirements has proven problematic in enlarging capabilities for sustainable development in Africa, primarily because of enormous pressure to adhere to goals and objectives determined by the donor without consultation.

Aid to Africa and Asia needs to be balanced or re-engineered to allow for rights and full ownership, as well as increased appropriate and localized strategic development interests. Perlette Louisy (2001, p. 436 quoted by Mason, 2011) argues that "Southern voices need to be taken seriously, not least for the very good reason of helping to prevent the uncritical and inappropriate transfer of policies from one context to another, probably very different context", which demonstrates mutual benefit and a Rights-Based Approach to aid. In the Rights-Based ideology, aid is significant only if it promotes ownership and control of the development process and enlarges opportunities for mutual respect that could be converted into that which serves the best interests of society. The tenet is the relevance of programming rights principles derived from these instruments – commitments to equality, inclusion, choice and voice (Babaci-Wilhite et al., 2012a). Poverty reduction is important but that there needs to be an equivalent focus on social inclusion participation and human rights, which leads us to question the effectiveness and efficiency of current aid architectures. With the target date for the MDGs only three years away, an estimated 1.4 billion people in the world are still living on less than 1.25 US dollars per day (World Bank, 2008).

Waun and Geo-JaJa (2013, p. 3) argue that ethically and in the context of social justice that knowledge production is not the monopoly of the North. How can the University of Oslo collaboration, based on Norad funding, contribute effectively to meeting development challenges and objectives in line with the national development strategies and plans of each African country involved? This is where the Rights-Based Approach in the context of the Norwegian aid paradigm comes into place. In order to meet the challenge of achieving sustainable development results, countries need the capacity to make the policy decisions that are right for them. Samoff (2003) claims that "with foreign funding came ideas and values, advice and directives on how education systems ought to be managed and targeted" (p. 440). Such aid construct is against the tenets of the Paris Declaration of 2005.

The lack of response of Northern universities to Southern interests constitutes a strong challenge in North–South research cooperation. For example, within the PITRO program the University of Oslo was not able to respond to the demand of the University of Dar-es-Salaam in the education sector due to a lack of interest on the part of the University of Oslo researchers. The interests of the North thus sometimes predominate over the interests of the South (OECD, 2005/2008), which is against the Rights-Based Approach of Norway's aid construct.

Monitoring the impact of university cooperation has also been a challenge. The focus of monitoring efforts needs to be broadened to include impacts on the living conditions of the beneficiaries of development aid, measuring whether lives have been changed in meaningful ways as indicated by the Nairobi outcome document, which states "The impact of South–South co-operation should be assessed with a view to improving, as appropriate, its quality in a results-oriented manner" (OECD, 2011, p. 4). In the same line, the Accra Agenda for Action states that the judgement will be made "by the impacts that our collective efforts have on the lives of poor people – rather than on the inputs and instruments – the focus of development co-operation has to be on delivering results" (OECD, 2011, p. 4). Collaborative North–South programs frequently lack a long-term perspective. PITRO will continue to operate under a different name, but the model will be maintained by supporting new projects rather than sustaining the PITRO project portfolio. The University of Oslo has not neglected the important matter of sustainability, as development cooperations often do.

Capacity-building and staff training will be achievable under the new PITRO and NORHED. The acknowledgment of the innovation and knowledge production that a society uses to define and shape its own development is crucial; therefore the research component needs to be developed in an equal partnership, as has been the case in PITRO. The program should have facilitated the publication of academic journal articles, which would strengthen the international ranking of the university. The voices of the South should be heard through publications in top-tier journals, which would help uphold the principles of the Rights-Based Approach of equality in inclusion and voice.

The alternative for effective collaboration in development is to forge new partnerships that, while accounting for different roles and responsibilities, rest on agreed common values (Chisholm & Steiner-Khamsi, 2009). This common ground translates into a new architecture for development of equal cooperation and recognition, where the South contributes as much as the North when it comes to knowledge building. The UN called for a mainstreaming of human rights to encourage governments' responsibility to insure the Rights-Based Approach. A Rights-Based Approach works to shift the paradigm away from aid and toward moral duty imposed on the world through the international consensus of human rights. South–South cooperation is growing due to the increasing acknowledgment among Southern countries that they share common goals. There is a dire need for new thinking about African cooperation with the North and its role in development – in short, for an educational transformation in Africa based on cooperation and sharing of expertise within the continent. This is in line with the 1997 Southern African Development

Committee (SADC) Protocol on Education (SADC, 1997), which prioritized sharing of expertise within the region. This sharing can facilitate cross-national learning and better decisions, acceleration of the pace of changes at lower costs and the most effective use of expertise on the continent. This in turn can bring about educational systems that are better suited to African learning environments and that are designed in the context of Africa's development needs. The chances of achieving these goals grow if Southern countries act together (Panitchpakdi, 2006). By joining efforts and building on common ground, development can be achieved on equal terms with dignity and can reach the people in need. However, authors such as Angeline Barrett, Michael Crossley and Hillary Dachi (2011) question to what extent ownership can genuinely be shared. While partners' agency in the North focuses on targets, outputs, incomes and outcomes, and deadlines, partners' agency in the South focuses on the process itself, on building capacity and building long-term relations with colleagues from the North.

To summarize, the overall findings of this chapter are that the Norwegian North–South university partnerships are successful to the degree that they are incorporating the Rights-Based Approach and giving importance to Southern voices. In order to achieve its goals, development collaboration with Africa has to be transformed to meet the internationally agreed MDGs within the Rights-Based Approach. As it stands now, these will not be reached by 2015; therefore this chapter suggests that strength in aid effectiveness will be accomplished by assigning priority to the following: (1) Country ownership, (2) Capacity-building and (3) Results which are sustainable. I agree with Mason (2011) when he writes that ownership should be accomplished with a common understanding of development goals and strategies and with a moral responsibility. The Rights-Based Approach provides an alternative means to address these issues.

Experts have evaluated the model of PITRO as a true partnership in which collaboration has on the whole met local needs. The universities involved have generated knowledge through research and disseminated that knowledge through teaching, both within the university but also to the local people involved in the projects. Through this model, the university makes the knowledge generated available to the society; therefore, strengthening of universities is key and fundamental for the development of the society as long as it is sustainable.

Collaboration between South and North must be envisioned in the context of Africa's interests, through broader efforts to achieve social, economic, and political rights in order to facilitate Africa's integration into the global economy now and in the future. The next steps within the Rights-Based Approach are to educate both the rights holders and the duty bearers by articulating the rights of citizens and the duty of the government. This is in line with a principle inherent in the Tanganyika African National Union (TANU) Arusha Declaration in the URT. It makes the explicit point that the inherent dignity of the individual should be respected, a point which is also in accordance with the Universal Declaration of Human Rights (UDHR) (Nyerere, 1967, p. 2).

To end this chapter I recommend that South–South university collaboration should be developed and incorporated in future development aid programs for sustainable development in Africa as well as in Asia.

6
Linguistic Rights for Appropriate Development in Education

Education in many African and Asian countries today has a range of serious harmful consequences, which violate various aspects of African and Asian linguistic rights in education and perpetuate poverty. Binding educational linguistic rights, including a right to MT-medium education in schools, is essential at primary level, and should be supplemented at the secondary and tertiary levels with good teaching of a local language as a second language. English should be taught as a foreign language. As it is, most children in Africa and Asia have to accept "subtractive" education through the medium of a foreign language. Children learn a dominant language at the cost of the MT, which is displaced, and later often replaced, by a foreign language. Teaching through a foreign language subtracts from the child's linguistic repertoire, instead of adding to it. Educational models for indigenous language in schools for children which use foreign languages as a LoI can and do have extremely negative consequences for the achievement of an effective learning process. In this chapter I provide the background on human rights conventions and argue that the use of local languages and local curricula in education should be regarded as a human right – a right in education.

The background of linguistic rights

The defining statement of human rights in the world is the UDHR (Table 6.1).

Table 6.1   Human rights conventions[1]

---

1. Convention on the Prevention and Punishment of the Crime of Genocide. Paris, December 9th, 1948.

2. International Convention on the Elimination of All Forms of Racial Discrimination. New York, March 7th, 1966.
   - Amendment to Article 8 of that Convention. New York, January 15th, 1992.

3. International Covenant on Economic, Social and Cultural Rights. New York, December 16th, 1966.
   - Optional Protocol to that Covenant. New York, December 10th, 2008.

4. International Covenant on Civil and Political Rights (ICCPR). New York, December 16th, 1966.
   - Optional Protocol to that Covenant. New York, December 16th, 1966.

5. Convention on the Non-Applicability of Statutory Limitations to War Crimes and Crimes against Humanity. New York, November 26th, 1968.

6. International Convention on the Suppression and Punishment of the Crime of Apartheid. New York, November 30th, 1973.

7. Convention on the Elimination of All Forms of Discrimination against Women. New York, December 18th, 1979.
   - Amendment to Article 20, Paragraph 1 of that Convention. New York, December 22nd, 1995.
   - Optional Protocol of the Convention. New York, October 6th, 1999.

8. Convention against Torture and Other Cruel, Inhuman and Degrading Treatment or Punishment. New York, December 10th, 1984.
   - Amendments to Articles 17 (7) and 18 (5) of that Convention. New York, September 8th, 1992.
   - Optional Protocol to that Convention. New York, December 18th, 2002.

9. International Convention against Apartheid in Sports. New York, December 10th, 1985.

10. Convention on the Rights of the Child. New York, November 20th, 1985.
    - Amendment to Article 43 (2) of that Convention. New York, December 12th, 1995.
    - Optional Protocol to that Convention on the Involvement of Children in Armed Conflict. New York, May 25th, 2000.
    - Optional Protocol to that Convention on the Sale of Children, Child Prostitution and Child Pornography. New York, May 25th, 2000.
    - Optional Protocol to that Convention on a Communications Procedure. New York, December 19th, 2011.

11. Second Optional Protocol to the ICCPR, Aiming at the Abolition of the Death Penalty. New York, December 15th, 1989.
12. International Convention on the Protection of the Rights of All Migrant Workers and Members of Their Families. New York, December 18th, 1990.
13. Agreement Establishing the Fund for the Development of the Indigenous Peoples of Latin America and the Caribbean. Madrid, July 24th, 1992.
14. Convention on the Rights of Persons with Disabilities. New York, December 13th, 2006.
    - Optional Protocol to that Convention. New York, December 13th, 2006.
15. International Convention for the Protection of All Persons from Enforced Disappearance. New York, December 20th, 2006.

Source: http://www.ohchr.org.

Eleanor Roosevelt was one of the pioneers in universalizing human rights. She chaired the UDHR Drafting Committee in 1945 and was recognized as a major contributor to the document. The UN Commission on Human Rights proceeded to draft two treaties:

(1) The International Covenant on Civil and Political Rights (ICCPR);
(2) The International Covenant on Economic, Social and Cultural Rights (ICESCR).

Together with the UDHR, they are commonly referred to as the International Bill of Human Rights.

The ICCPR focuses on issues such as the right to life and freedom of speech, religion and voting. The ICESCR focuses on issues such as food, education, health and shelter. Both covenants trumpet the extension of rights to all persons and prohibit discrimination.

The UDHR (1948) was proclaimed as a common standard of achievement for all people and all nations. This Declaration strives, by teaching and educating, to promote respect for these rights and freedoms, national and international, to secure their universal and effective recognition and observance among the peoples. Eleanor Roosevelt remarked at the UN on March 27th, 1958:

Where after all do universal human rights begin? In small places, close to home – so close and so small that they cannot be seen on any map of the world. Such are the places where every man, woman, and child seeks equal justice, equal opportunity, equal dignity without discrimination. Unless these rights have meaning there, they have little meaning anywhere. Without concerted citizen action to uphold them close to home, we shall look in vain for progress in the larger world.

The UDHR sets out a list of basic rights for everyone in the world whatever their race, color, sex, language, religion, political or other opinion, national or social origin, property, birth or other status. It states that governments have promised to uphold certain rights, not only for their own citizens, but also for people in other countries. The UDHR was unanimously adopted on December 10th, 1948 by the UN (although eight nations did abstain). In 1993 a world conference of 171 states representing 99% of the world's population reaffirmed its commitment to human rights (OHCHR, 1996–2014).

According to the UDHR, everyone has the right to education. Education shall be free, at least in the elementary and fundamental stages. Elementary education shall also be compulsory. Technical and professional education shall be made generally available and higher education shall be equally accessible to all on the basis of merit. Education shall be directed to the full development of the human personality and to the strengthening of respect for human rights and fundamental freedoms. It shall promote understanding, tolerance and friendship among all nations and all racial or religious groups, and shall further the activities of the UN for the maintenance of peace.

Parents have a prior right to choose the kind of education that shall be given to their children. In Article 13, paragraph 1 of the ICESCR, states parties also agree that *education* shall be directed to the full development of the human personality and the sense of its dignity, and shall strengthen respect for human rights and fundamental freedoms. Combined with the comments just made with respect to Article 13, paragraph 1 of the ICESCR, it would seem clear that an education in a language other than the child's MT shows disrespect for the child's cultural identity, language and values (Table 6.2).

The human rights proclaimed in the UDHR are inscribed in documents such as the International Covenants on Human Rights, which sets out what governments must do and also what they

Table 6.2   Universal Declaration of Human Rights[2]

Universal Declaration:	Universal Declaration:
Education shall be free, at least in the elementary and fundamental stages. Elementary education shall be compulsory.	Parents have a prior right to choose the kind of education that shall be given to their children.
UNESCO Convention against Discrimination in Education:	UNESCO Convention against Discrimination in Education:
The States Parties to this Convention undertake to formulate, develop and apply a national policy which ... will tend to promote equality of opportunity and of treatment ... and in particular: (a) To make primary education free and compulsory.	The States Parties to this Convention agree that: (b) It is essential to respect the liberty of parents ... firstly to choose for their children institutions other than those maintained by the public authorities but conforming to ... minimum educational standards, and secondly, to ensure ... the religious and moral education of the children in conformity with their own convictions.
International Covenant on Economic, Social and Cultural Rights:	International Covenant on Economic, Social and Cultural Rights:
Primary education shall be compulsory and available free for all.	The States Parties to the present Covenant undertake to have respect for the liberty of parents ... to choose for their children schools, other than those established by the public authorities, which conform to such minimum educational standards as may be laid down or approved by the State and to ensure the religious and moral education of their children in conformity with their own convictions. No part of this article shall be construed so as to interfere with the liberty of individuals and bodies to establish and direct educational institutions ...

Table 6.2 (Continued)

Convention on the Rights of the Child: States Parties recognize the right of the child to education, and with a view to achieving this right progressively and on the basis of equal opportunity, they shall, in particular: (a) Make primary education compulsory and available free for all.	Convention on the Rights of the Child: No part of [Articles 28 and 29] shall be construed so as to interfere with the liberty of individuals and bodies to establish and direct educational institutions ...

must not do to respect the rights of their citizens. It is important to note that human rights do not have to be given, bought, earned or inherited, they belong to people simply because they are human:

- Human rights are *inherent* to each individual, they cannot be taken away;
- We are all born free and equal in dignity and human rights are *universal* because they apply to everyone in the world;
- People still have human rights even when the laws of their countries do not recognize them, or when they violate them; human rights are *inalienable*;
- To live in dignity, all human beings are entitled to freedom, security and decent standards of living concurrently – human rights are *indivisible*.

Rights can be put into three categories:

(1) Civil and political rights. These are "liberty-orientated" and include the rights to: life, *liberty* and security of the individual; freedom from torture and slavery; political participation; freedom of opinion, expression, thought, conscience and religion; freedom of association and assembly.
(2) Economic and social rights. These are "security-orientated" rights, for example the rights to: work, *education*, a reasonable standard of living, food, shelter and health care.

(3) Environmental, cultural and developmental rights. These include the right to live in an environment that is clean and protected from destruction, and rights to *cultural,* political and economic development.

The United Nations Convention on the Rights of the Child (CRC) was written in 1989 and came into force in 1990. All the countries in the world have agreed to it except the USA, South Sudan and Somalia. The CRC held in 2003 recommended that:

(1) States ensure access to appropriate and high-quality education. This is a necessary prerequisite for preparing children for higher education.
(2) The educational provisions of the United Nations General Assembly Declaration of the Rights of Persons belonging to National or Ethnic, Religious and Linguistic Minorities of 1992, article 4, paragraph 3 provides that: "States should take appropriate measures so that, wherever possible, persons belonging to minorities have adequate opportunities to learn their mother tongue or to have instruction in their mother tongue".

(UN, 1948)

Article 28, paragraph 1 of the CRC recognizes the right of the child to education; significantly, the paragraph proceeds to specify that States shall, in particular: "take measures to encourage regular attendance at schools and the reduction of drop-out rates" and Article 29, paragraph 1 of the CRC states that (Table 6.3):

Table 6.3 Global human rights standards regarding language[3]

UNESCO Convention against Discrimination in Education (1960)
It is essential to recognize the right of members of national minorities to carry out their own educational activities, including the maintenance of schools and depending on the educational policy of each state, the use or the teaching of their own language, provided however:
   That this right is not exercised in a manner which prevents the members of these minorities from understanding the culture and

---

74

Table 6.3 (Continued)

the language of the community as a whole and from participating in its activities, or which prejudices national sovereignty;
    That attendance at such schools is optional.

International Covenant on Civil and Political Rights (1966)
In those States in which ethnic, religious or linguistic minorities exist, persons belonging to such minorities shall not be denied the right, in community with other members of their group, to enjoy their own culture, to profess and practice their own religion, or to use their own language.

ILO Indigenous and Tribal Peoples Convention (1989)
Measures shall be taken to ensure that members of the (indigenous and tribal) peoples have the opportunity to acquire education at all levels on at least an equal footing with the rest of the national community. Education programmes and services for the (indigenous and tribal) peoples shall be developed and implemented in co-operation with them...
    In addition, governments should recognize the right of these peoples to establish their own educational institutions and facilities, provided that such institutions meet minimum standards established by the competent authority in consultation with these peoples. Appropriate resources shall be provided for this purpose.
    Children belonging to the (indigenous and tribal) peoples shall, whenever practicable, be taught to read and write in their own language or in the language most commonly used by the group to which they belong. When this is not practicable, the competent authorities shall undertake consultations with these peoples with a view to the adoption of measures to achieve this objective.
    Adequate measures shall be taken to ensure that these peoples have the opportunity to attain fluency in the national language or in one of the official languages of the country.

Convention on the Rights of the Child (1989)
In those States in which ethnic, religious or linguistic minorities or persons of indigenous origin exist, a child belonging to such a minority or who is indigenous shall not be denied the right, in community with other members of his or her group, to enjoy his or her own culture, to profess and practice his or her religion, or to use his or her own language.

The United Nations Permanent Forum on Indigenous Issues examined indigenous and minority education and concluded that the length of mother tongue LoI is more important than any other factor, including socio-economic status, in predicting the educational success of students. The worst results, including high push-out rates, as Africa and Asia have had for several decades, are associated with students in programs in which the students' indigenous languages are not supported or where they are only taught as subjects.

The results of the UN study are in line with the proposition of economist Nobel Prize laureate Amartya Sen, that poverty is not only about economic conditions and growth; expansion of human capabilities is a more basic remedy for poverty and more basic objective of development. Dominant foreign LoI for children curtails the development of their capabilities and perpetuates poverty. The present practices of educating children through the medium of dominant languages are completely contrary to both solid theories and research results, such as those presented in my edited volume *Giving Space to African Voices: Rights in Local Languages and Local Curriculum* (2014). The book examines African projects and Asian cases to find how best to achieve the goals of quality education and secure the rights to education that children have according to international law.

It is important to understand the distinction between rights in education and right to education. Right to education is access to schooling, whereas Babaci-Wilhite and Geo-JaJa (2014) define rights in education as access to indigenous knowledge in indigenous language. With regard to the right to education, at least two major UN instruments create binding obligations for participating states to recognize the right of everyone to access education.

In a very important report on the rights of children launched in 2004, UNICEF states that "Illiteracy is a direct result of educational exclusion". The effects of enforced foreign dominant LoI policies, which tend to result not only in considerably poorer performance but also higher levels of non-completion and educational exclusion, could be said to be contrary to sub-paragraph 1 of Article 28. Research on educational performance indicates that children from indigenous linguistic backgrounds taught through a foreign language perform considerably worse than native dominant language-speaking children in the same class, and that they suffer from higher levels of push-out rates, as demonstrated in Tanzania in 2013.

The 4As: Availability, Accessibility, Acceptability and Adaptability

The legal basis for education, as stated by the former UN Special Rapporteur on the Right to Education Katarina Tomasevski (2006), is that linguistic rights can be divided into individual and collective rights. Tomasevski (2003) advocates that education should prepare learners for participation: "it should teach the young that all human beings – themselves included – have rights" (2003, p. 33). Furthermore, Tomasevski argues that this dominant LoI prevents access to education because of the linguistic, pedagogical and psychological barriers it creates. Regarding the purpose of education, she questions whether it is reinforcing or eliminating inequality.

Tomasevski states that "mere access to educational institutions, difficult as it may be to achieve in practice, does not amount to the right to education" (2004, paragraph 57). The first feature she mentions as important is the use of the official local language of the country as the LoI in school. If the educational model chosen for a school legally or administratively does not mandate or even allow children to be educated mainly through the LoI that the child understands, then the child is effectively being denied access to education. If the teaching language is foreign to the child and the teacher is not properly trained to make the subject comprehensible in the foreign language, the child does not have access to quality education. Likewise, if the LoI is neither the MT/first language nor minimally an extremely well-known second language, the child does not have equal access to good education. Here, according to Tomasevski, there is a combination of linguistic, pedagogical and psychological barriers to "access" to quality education. She argues that there is the need for a right to quality education and that these should include what she refers to as the 4-A Scheme: Availability, Accessibility, Acceptability and Adaptability, as explained in the table below (Table 6.4):

Table 6.4 Conceptual framework[4]

Right to education	Availability	– Fiscal allocations matching human rights obligations – Schools matching school-aged children (number, diversity) – Teachers (education & training, recruitment, labor rights, trade union freedoms)
	Accessibility	– Elimination of legal and administrative barriers – Elimination of financial obstacles – Identification and elimination of discriminatory denials of access – Elimination of obstacles to compulsory schooling (fees, distance, schedule)
Rights in education	Acceptability	– Parental choice of education for their children (with human rights correctives) – Enforcement of minimal standards (quality, safety, environmental, health) – LoI – Freedom from censorship – Recognition of children as subjects of rights
	Adaptability	– Minority children – Indigenous children – Working children – Children with disabilities – Child migrants, travellers
Rights through education		– Concordance of age-determined rights – Elimination of child marriage – Elimination of child labor – Prevention of child soldiering

Availability embodies two different governmental obligations: the right to education as a civil and political right requires the government to permit the establishment of educational institutions by non-state actors, while the right to education as a social and economic right requires the government to establish them, or fund them, or use a combination of these and other means so as to ensure that education is available.

Accessibility is defined differently for different levels of education. The government is obliged to secure access to education for all children in the compulsory education age range, but not for secondary and higher education. Moreover, compulsory education ought to be free of charge while post-compulsory education may entail the payment of tuition and other charges and could thus be subsumed under "affordability". The increasing trend of charging fees for post-compulsory education is contrary to the spirit of international human rights law.

Acceptability of education has been highlighted by the addition of the modifier "quality" education in policy documents from the 1990s, thus urging governments to ensure that the education that is available and accessible is of good quality. The minimal standards of health and safety, or professional requirements for teachers, thus have to be set and enforced by the government. The scope of acceptability has been considerably broadened through the development of international human rights law. Censorship of school textbooks is no different from any other censorship, except in that it is infrequently exposed as a human rights violation. The focus on indigenous and minority rights has prioritized the LoI, which often makes education unacceptable if the language is foreign to young children (and also often to the teacher). The prohibition of corporal punishment has transformed school discipline in many countries, further broadening the criteria of acceptability. The emergence of children themselves as actors seizing their right to education and rights in education promises to endow the notion of acceptability with their vision of how their rights should be interpreted and applied.

Adaptability has been best conceptualized through the many court cases addressing the right to education of children with disabilities. Domestic courts have uniformly held that schools ought to adapt to children, following the thrust of the idea of the best interests of each child in the Convention on the Rights of the Child. This reconceptualization has implicitly forced children to adapt to whatever schools that may have been made available to them; the school effectively had a right to reject a child who did not fit or could not adapt. Moreover, a conceptual dissociation between "school" and "education" has taken place in attempts to provide education to imprisoned or

Table 6.5 The 4-A Scheme[5]

Availability	Schools	Establishment/closure of schools
		Freedom to establish schools
		Funding for public schools
		Public funding for private schools
	Teachers	Criteria for recruitment
		Fitness for teaching
		Labour rights
		Trade union freedoms
		Professional responsibilities
		Academic freedom
Accessibility	Compulsory	All-encompassing
		Free of charge
		Assured attendance
		Parental freedom of choice
	Post-compulsory	Discriminatory denials of access
		Preferential access
		Criteria for admission
		Recognition of foreign diplomas
Acceptability	Regulation and supervision	Minimum standards
		Respect for diversity
		LoI
		Orientation and contents
		School discipline
		Rights of learners
Adaptability	Special needs Out-of-school education	Children with disabilities
		Working children
		Refugee children
		Children deprived of their liberty

working children. They can seldom be taken to school and thus education has to be taken to wherever they are (Tomasevski, 2006, pp. 13, 15) (Table 6.5).

Language in education as a human right

All education as defined by recognized international human rights standards should teach about and advocate for human rights by:

(1) Teaching people about their rights and responsibilities;
(2) Teaching people how to respect and protect rights.

In this context Amnesty International defines human rights education as a program which aims to provide knowledge and understanding about human rights, and seeks to introduce human rights values in the teaching or training practices and curricula of both formal and non-formal educational programs. Formal education is understood as the official education system comprising nursery, primary, secondary and tertiary education. Non-formal education refers to those teaching programs outside the formal education system, often managed by Non-Government Organizations (NGOs), which aim to provide literacy and other skills to the many millions of children and adults who are denied access to the formal education system.

Education about and for human rights includes the development of skills such as critical thinking, communication skills, problem-solving and negotiation, all of which are essential for effective human rights activism and participation in decision-making processes. Human rights education is all about helping educators/teachers/trainers to understand human rights and to feel that these are important and should be respected, defended and taught to all students everywhere regardless of age, gender, ethnic background or the educational setting. Teaching for and about human rights involves the use of participatory methodology. Participatory, interactive methodology has been found by educators to be the most relevant and appropriate way to develop skills and attitudes, as well as knowledge, in both children and adults. Such methodology involves students fully in their own learning. They become active explorers of the world around them, rather than passive recipients of the educator's expertise. This methodology is particularly appropriate when dealing with human rights issues, where there are frequently many different points of view on an issue, rather than one "correct" answer.

There is a need for national education policies grounded in the Rights-Based Approach since education is not only a right in itself but also "indispensable for the exercise of other human rights". A paradigm shift of the present practices of educating children through contextualized curriculum and the medium of foreign/official/state languages is crucial and a linguistic right. Article 26 of the UDHR (1948) states that:

(1) Everyone has the right to education. Education shall be free, at least in the elementary and fundamental stages. Elementary

education shall be compulsory. Technical and professional education shall be made generally available and higher education shall be equally accessible to all on the basis of merit.

(2) Education shall be directed to the full development of the human personality and to the strengthening of respect for human rights and fundamental freedoms. It shall promote understanding, tolerance and friendship among all nations, racial or religious groups, and shall further the activities of the UN for the maintenance of peace.

(3) Parents have a prior right to choose the kind of education that shall be given to their children.

This says little about the nature, kind and quality of education. There is a need to bring to the discussion of educational rights the notion of rights in education, which implies that rights are not ensured unless the education offered is of high quality.

Rights and capabilities are often discussed as multi-dimensional models, which can be seen as comprehensive models. The UN called for a mainstreaming of human rights to encourage the government's responsibility to insure the Rights-Based Approach. The Rights-Based Approach works to shift the paradigm toward moral duty imposed on the world through the international consensus of human rights. The Rights-Based framework includes the principle that every human being is entitled to decent education and gives priority to the intrinsic importance of education, implying that governments need to mobilize sufficient resources to offer quality education (UNICEF, 2003, p. 8).

Rights in education should embrace protection of and respect for learners' cultures, needs and languages (Babaci-Wilhite et al., 2012). Individuals have the right to use their MT in their private lives and in their social and cultural activities, and to transmit the MT to their children through education. Collective linguistic rights are related to cultural and group identity, which must be valorized and preserved. It is in education that linguistic human rights are most often violated. In Article 2 of the UDHR (1948), it is stated that:

Every person has the right to enjoy human rights regardless of their race, color, sex, language, religion, or opinion.

In Article 26.1, it is stated that:

> Every person has the right to education; the latter must be free; at least fundamental education (primary and secondary), and primary education must be compulsory.

Robeyns (2006) claims that the rights-based discourse runs the risk of reducing rights to legal rights only. She points to Thomas Pogge (2002, pp. 52–53), who argues that human rights can be understood as moral rights or as legal rights, which can in principal co-exist and can be complementary. However, he writes that a weakness of this inclusive view is that human rights are whatever governments agree for them to be. Robeyns (2006) argues, "the rights-based approach model of education is that, once the government agrees that children should have the rights to be educated, it may see its task as being precisely executing this agreement, and nothing more". Furthermore, she claims that "well-developed rights-based educational policies will state precisely which rights are guaranteed to whom, and what the government has to do to ensure that rights are not only rhetorical, but also effective" (ibid., p. 77). Education has the potential to empower if teaching and learning provide nourishment and self-respect that in turn bring confidence to teachers and learners (Spreen & Vally, 2006).

Although the UDHR states that "Parents have a prior right to choose the kind of education that shall be given to their children", in several Asian and most African countries many parents want their children to learn not in local languages but in English; even if they themselves do not speak English, they want English as a LoI for their children. Despite the evidence that a child learns best in his/her MT, most parents do not understand the implications for their children of learning in a language that is not their own (Babaci-Wilhite, 2010). One of the important findings from my research is that parental choices between public versus private schools and between local languages and English are made on the basis of imperfect information about the learning implications of these choices. The private schools are doing better because they have more resources than government schools (Babaci-Wilhite, 2010; Bakahwemama, 2010; John, 2010).

It is essential that the government provide better information on the role of language in learning and on the advantages of local languages as LoI. When confronted with this, government officials respond that it should be the parents' responsibility to seek out this information, and that the government, as a democracy, should respect parental choices. However, my research indicates that the parental misunderstanding about language and learning is based on a myth: Parents believe that having English, as the LoI will improve students' learning abilities and their opportunities in life. The myth has to be deflated in order for parents to make informed choices (Babaci-Wilhite, 2010).

The famous Tanzanian NGO Haki Elimu (Rights for Education) disseminated a short film in Tanzania entitled *HakiElimu Lugha* (Haki Elimu, 2009) to teach parents the consequences of using English as a medium of learning in schools, but it did not affect parents' preference for having English in school. The parents misunderstood the film, and have even said that it is a "plot" by the government, which really wants to keep their children uneducated. Some of the parents interviewed wondered why government officials were sending their own children to English private schools if they believed that Kiswahili is good for the learning process. But they argued that the government neglects government schools and government officers send their children to English-medium schools, sometimes abroad. One of my interviewees said, "They want their children to learn English because they have an advantage, better power, equipment, I will say it is political."

I agree with Robeyns (2006, p. 77) when she writes that "It will be necessary that the government goes beyond its duties in terms of the rights-based policies, to undertake action to ensure that every child can fully and equally enjoy her rights to education." This implies that teachers must be well trained and well paid, and that adequate teaching materials must be provided and a good curriculum and pedagogy developed.

As argued and reviewed in Babaci-Wilhite (2013a), the choice of the LoI is crucial for learning. Language plays a critical role in cognitive learning and in the development of logic, reason, critical thinking and new knowledge (Lwaitama, 2004; Brock-Utne, 2007; Babaci-Wilhite, 2013a; Bostad, 2013) and should be seen as a human right. UNESCO's convention on the Protection and Promotion of the Diversity of Cultural Expressions emphasizes the importance of linguistic diversity (2005). Local language should be seen as an intimate part of culture and thus should be designated as a human right in the education sector (Skutnabb-Kangas, 2000). Applying the arguments regarding quality learning and capability approach, education in a local language should be regarded as a human right.

Reforms and policies connecting local cultures to education have been neglected in Africa. According to Samoff (2007, p. 60), "effective education reform requires agendas and initiatives with strong local roots". In other words, indigenous knowledge should be included in the curriculum (Odora, 2002; Breidlid, 2009), and indigenous language is critical to the preservation and development of indigenous knowledge. Many educational practitioners continue to ignore culture as a central ingredient in education. Geo-JaJa (2009, p. 93) affirms that "the alien factory model of schooling 'Western' educational system that is rooted in mechanistic and linear worldview that is found in most developing countries today that oppose traditional values are inconsistence to right in education".

With the notion behind EFA, many educational practitioners continue to ignore intrinsic factors, particularly culture, as a central core of education. Moreover, this is more significant as the needs of rural excluded communities are rarely captured in school reforms or policy, including those directed at the poor or those located in more isolated areas. The use in schooling of a local language that students are familiar with significantly redistributes access to quality education among the elites and the masses and also strengthens African languages, to the detriment of hegemonic forces promoting the use of colonial languages (Babaci-Wilhite, 2014a).

There are political and global influences promoting English, since English is the global language and the most important language in the world. However, during my research in Africa, it has been highlighted that there has been a long debate over the change to English and that a change now is possibly a bad choice. One of the leaders ended with the following statement:

> We are so much influenced by our colonialism mentality, we forget that our children can think and process a lot more things through their mother tongue than in English most of them are stuck and can not go to higher education I am sure it is the reason because they are very intelligent.

However, one of my interviewees said, "That is true that the elite favour English as a language of instruction and that is a problem. I am myself repeating why, and again I repeat what Franz Fanon called this mental colonization must stop."

To summarize, African and Asian countries will not achieve human rights in education until and unless they acknowledge that local language, identity and culture are to be respected and fulfilled in local curricula. Teaching a child in a language s/he does not master disrespects the child's cultural identity, language and values and goes against Article 13, paragraph 1 of the ICESCR, and the Committee on the Rights of the Child, held in 2003, in which states ensure access to appropriate and high-quality education. Furthermore, the educational provisions of the United Nations General Assembly Declaration of the Rights of Persons belonging to National or Ethnic, Religious and Linguistic Minorities of 1992, Article 4, paragraph 3 affirm that "States should take appropriate measures so that, wherever possible, persons belonging to minorities have adequate opportunities to learn their mother tongue or to have instruction in their mother tongue" (UN, 1948). This is not the case in Africa and in many Asian countries.

If the educational model chosen for a school legally or administratively does not mandate or even allow children to be educated mainly through the LoI that the child understands, then the child is effectively being denied their human rights.

7

Language-in-Education Policy

The focus in this chapter is on two comparisons of the role of LoI in language-in-education policy, first of all comparing Tanzania and Nigeria and secondly comparing Zanzibar with Malaysia. All of these countries have a British colonial legacy and a history characterized by debates about language politics and the choice of language in education. It is important to mention that Tanzania and Zanzibar are one country. Tanganyika gained its independence from Britain on December 9th, 1961 and Zanzibar was declared independent on December 10th, 1963. Tanganyika and Zanzibar united on April 26th, 1964 to form the URT, however, the Revolutionary Government of Zanzibar has considerable autonomy over its internal affairs administratively, and has its own legislative body and executive functions including its own ministries.

The policies of Tanzania and Nigeria lend themselves to comparison because both countries adopted African languages as a LoI decades ago. Tanzania has one major African language, Kiswahili, and uses it as a LoI in all years of primary school. Nigeria has three major African languages that are used as LoI in primary schools. Hausa is a transnational language, while Igbo and Yoruba are regional languages for their respective regions.

The rationale for comparing Zanzibar with Malaysia is recent curriculum changes that are diametrically opposed in their choice of LoI. Zanzibar is replacing Kiswahili with English as a LoI in mathematics and science from Grade 5, while Malaysia recently reversed a decision in 2003 to replace Bahasa Malaysia (the Malay language) with English for mathematics and science subjects and reinstated Bahasa Malaysia as the LoI in 2010 (Gill, 2005).

A comparative study of Tanzania and Nigeria

Tanzania is one of the few countries in Africa that has an African national language, Kiswahili (Figure 7.1).

The efforts to promote Kiswahili in Tanzania began as early as the 1930s. President Julius Nyerere had a strong vision of education, which he regarded as an important contributor to social action. He believed that Kiswahili should be the Tanzanian language and the language used in the educational system. He was successful in making Kiswahili the official national language and the language used in primary education. The example of Tanzania demonstrates that African languages have the same potential to serve as a pan-national language as any other language, and language unification can be made to happen if there is political will to create and enforce the necessary policies and strategies.

In Tanzania, the National Kiswahili Council, in Kiswahili Bakita (Baraza la Kiswahili la Taifa), was founded in 1967 by a government act. It was given a budget and a staff with the mandate to develop

Figure 7.1   A school in Tanzania

Kiswahili and to make sure the language was properly used in the media. After independence, the work of promoting the language was continued at the Institute of Kiswahili Research at the University of Dar-es-Salaam. In 1967, the Tanzanian constitution was amended and Kiswahili became formalized as the LoI for primary school grades within the education system. Kiswahili has since been used in Tanzania as both official and national language. A competence in English is also important, since English links Tanzania and the rest of the world as the global language of technology, commerce and administration (URT, 2009). Even so, in most official and legal discussions, Kiswahili is the language of choice. Haroub Othman, a Zanzibari scholar, gives us an example that conveys the tension between English and Kiswahili within the legal system:

> I remember an incident in 2007 at the General Meeting of the Zanzibar Law Society where members argued for some time whether the meeting should be conducted in English or Kiswahili. Later the President of the Society ruled that it should be in English since it was the official language of the High Court. Half an hour after the decision was made, nobody was talking in English, and no one protested.
>
> (2008, p. 6)

Kiswahili is often used as the intra-family language after marriage in Tanzania. About 80 million people in 14 countries in East and Central Africa speak Kiswahili (URT, 2009). Kiswahili is a language widely spoken in eastern Africa and adjacent islands, but also in other parts of Africa and Arabia, and is taught in many institutions of learning in Europe, Japan, Korea, the USA, the UK, Canada and elsewhere (Ismail, 2013). In Zanzibar, Kiswahili is understood by the entire population and spoken as MT by the vast majority. The language of wider communication at all levels is Kiswahili. English is a foreign language that was introduced in Tanzania (then Tanganyika) during British colonial rule. In spite of long exposure to English, today only 5% of the population speaks English (Kimizi, 2012).

Kiswahili was acknowledged as the language of choice beyond Tanzania's borders, becoming the official language for the inter-territorial East African Language Committee in Tanganyika

(Tanzania), Kenya and Uganda.[1] Kiswahili is one of the five official languages of the African Union alongside English, French, Portuguese and Arabic. Kiswahili has occasionally been used as a working language in United Nations Educational, Scientific and Cultural Organization (UNESCO) meetings as far back as 1986. It was, however, never made an official working language of the UN or UNESCO. Othman argues that:

> Kiswahili is no longer the language of Tanzania or East Africa; it is the language of the entire African continent, having been adopted by the African Union as one of its official languages. When former Mozambican President, Joaquim Chissano (and not the President of Tanzania, Benjamin Mkapa), addressed the African Heads of State Summit for the first time using Kiswahili, the audience warmly applauded.
>
> (2008, p. 7)

Language plays a major role in Tanzania's robust media. Most newspapers in Tanzania are in Kiswahili. The public broadcasting television service Televisheni ya Taifa, or Tanzania Broadcasting Cooperation, produces most of its programs in Kiswahili. The radio networks of Radio Tanzania Dar-es-Salaam are also state-run and use Kiswahili. It is important to note that from 2007 the Tanzanian state has owned both Televisheni ya Taifa and Radio Tanzania Dar-es-Salaam. However, privately owned media companies are more important, since they control more than 11 daily newspapers, more than six television stations and more than six FM radio stations. All of these are published or conducted in Kiswahili. One of them, Radio Free Africa, reaches the Great Lakes Region – the Democratic Republic of Congo, Rwanda and even Burundi. This shows the importance of Kiswahili as a cross-border language (URT, 2009).

The Kiswahili–English language dilemma remains a subject of intense debate among language and education scholars. Othman formulates the central question this way:

> Why is a country like Tanzania, which was in the forefront of Africa's liberation struggle, which proclaimed the Arusha Declaration that ushered in its own development path and which in its policy documents and proclamations wanted the people to be

the masters of their own destiny, unable to resolve this language problem?

(2008, p. 6)

Ouane and Glanz (2006) wrote that Tanzania, in comparison to Benin, Burkina Faso, Cameroon, Mali and Zambia, was the only one that went beyond experimentation and implemented a policy that promoted the effective use of a national language in formal and non-formal education and administration. As noted above, its success can be traced back to Nyerere's concept of "education for self-reliance", also known as Ujumaa, in the Arusha Declaration (1968), which perceives education as the means for laying the foundations in the present for future development (Figure 7.2).

In Nigeria, formal education in a Nigerian language began in 1831 with Yoruba in the Western Region of Nigeria; however, Nigerian languages are not used as LoI in post-primary education. Emmanuel Nwanolue Emenanjo (2008) writes that despite the intention of the 1926 Education Edict of the Colonial Government, supported by many noble sentiments of indigenous federal and state governments,

Figure 7.2   A school in Nigeria

and valiant activities of the Linguistic Association of Nigeria (ibid.), the use of Nigerian LoI has not advanced beyond the primary school level.

In 1970, Nigeria started a project entitled the "Six-Year Primary Project" (SYPP), based at the prestigious Institute of Education at the University of Ife (now Obafemi Awolowo University). The project was conducted in a rural school, setting up two experimental classes and one control class. In the control class, students used Yoruba as LoI throughout the six years of primary education. The results of this project have confirmed that those who receive their total primary education in their MT have proven to be more resourceful and academically better prepared. "The SYPP children have demonstrated greater manipulative ability in their relationship to their colleagues, they also tend to demonstrate a great sense of maturity, tolerance and other affective qualities that make them integrate easily and readily with those they come in contact with" (Fafunwa et al., 1989, p. 141).

In terms of pure academic attestation, results consistently showed that the experimental group performed best on tests of all subjects, including English. The research of the University of Ife SYPP shows that people learn faster and better in their first language. Clearly this outcome demonstrates that language policy is a significant contributor to quality learning in an educational system. Further, the impact is still underestimated despite research that proves the correct choice of language to be essential.

Consequently, to achieve functional literacy, it is imperative for Nigerians and other African students to be re-educated on the place of language in the teaching and learning process, as well as for the curriculum to be reformed to be consistent with national needs in their system of education. For example, the SYPP enabled a strengthening of the learning process and also enriched the curriculum by developing materials in Yoruba. This two-pronged process of education formulation resulted in more effective teaching of English as a subject through the use of specialist teachers of English (Geo-JaJa, 2009).

Nigeria has a Language Development Centre called the Nigerian Educational Research and Development Council. At this Centre a large body of scientific literature has been developed in MT in order to facilitate the use of the appropriate scientific words, terms and phrases. The 1989 Constitution of the Federal Republic of Nigeria states that the federal, state and local governments should offer scholarships to students of languages. The in-service training for teachers of languages in tertiary institutions should continue through the postgraduate level, and tertiary institutions should train language teachers through staff development programs and exchange of teachers between states and between institutions. These policies reinforcing teacher competence in local languages were only paid lip service by the government. Bamgbose (1982) has correctly identified the many barriers against effective education in West African languages in general and Nigerian languages in particular. Emananjo (2008) argues that a key constraint is an assumption that literacy is the ability to speak and write English, a fallacy that is strongest among the Southern Nigerian elite. However, after over 200 years of English in Nigeria, less than 20% of Nigerians are able to speak and write English (ibid.).

Bamgbose (1982) posits that if children were educated in an indigenous language, this would satisfy both the letter and the spirit of Section 1, Paragraph 8 of the Nigerian Policy of Education, which states:

> In addition to appreciating the importance of language in the educational process, and as a means of preserving the people's culture, the Government considers it to be in the interest of national unity that each child should be encouraged to learn one of the three major languages other than his own mother tongue. In this connection, the Government considers the three major languages in Nigeria to be Hausa, Igbo, and Yoruba.

The Nigerian Policy of Education (1977, revised in 1982, 1998, 2004) is the educational blueprint of the Federal Government of Nigeria. It stipulates the government policy on which languages should be used to educate children in schools. In Section 1, Paragraph 10 states:

> Government appreciates the importance of language as a means of promoting social interaction and national cohesion; and preserving cultures. Thus every child shall learn the language of the immediate environment. Furthermore, in the interest of national unity it is expedient that every child shall be required to learn one of the three Nigerian languages: Hausa, Igbo and Yoruba.

As argued by Ogunsuji (2003), the language policy is a binding language "guide" meant to be "enforced" by the society that formulated it through a political process and made operational. Thus it should be implemented with all the facilities necessary to achieve quality education, which is not the case in Nigeria according to Geo-JaJa and Azaiki (2010). Hausa, Igbo and Yoruba are major languages because they are spoken in at least five to ten states each and have a large numbers of speakers (Akande & Salami, 2010). In a report by Rafiu (2012), the author acknowledge the death of many languages in Nigeria such as: Ake (Nasarawa state), Bakpakia (Cross River), Butanci, Shau and Kudu-Camo (Bauchi), Chamba (Taraba state), Sheni (Kaduna state), Holma and Honta (Adamawa state) and Sorko languages (Niger, Kwhar and Kebbi states).

The local non-dominant languages should have orthographies to ensure the readership levels and should be used in post-primary and primary education. The local dominant languages such as Yoruba, Igbo and Hausa should be used in secondary schools. This would contribute to the power and prestige of these indigenous languages as recommended by Rafiu (2012) that politicians and union leaders should read their manifestoes in the languages of their people either in addition, or as an alternative, to doing so in European languages. This would help to change the negative and ambivalent attitudes of Nigerians toward the use of the languages and guarantee their viability.

In Nigeria, many children are quiet, learn little or nothing in class and fail in the end because they are not able to interact in a viable way in English. Egbe et al. (2004) argue that the use of English has an enormous impact on other aspects of life and is especially detrimental to achievements in the field of education. With the use of English, performance is lower than it should be, productivity is low and social interactions are inhibited. Furthermore, Omojuwa (2005) states that the majority of children in Nigeria are ill-prepared for the demands of academic studies by the time they enter tertiary institutions, as they depend on shortcuts to knowledge.

The idea of forcing students to think in a foreign language as advocated by Dikshit (1974) is unproductive. Olarenwaju (2008) states that this "does not help students to be creative but reduces them to 'robots' who merely memorize the notes given to them by their teachers and reproduce them when required without demonstrating

appreciable degree of understanding of the scientific and technological information and process under consideration". He furthermore
notes that if "students merely memorize facts, principles and gener-
alizations only to be regurgitated during examination, they will not
be in a position to use the knowledge acquired since it has not been
internalized". He concludes that the lack of internalization of scientific knowledge, process and skills by Nigerian students seems to
have been largely responsible for Nigeria's inability to make a major
breakthrough in scientific and technological development. Fafunwa
(1990) points out that the imposed LoI is an important factor mitigat-
ing against the dissemination of knowledge and skills, and therefore
directly impacts the rapid social and economic well-being of the
majority of people in Africa. There tends to be a correlation between
slower assimilation and the use of a foreign language as the official
language of a given country in Africa. No society in the world has
developed in a sustained and democratic fashion on the basis of a
borrowed or colonial language.

As discussed previously, it has been difficult for Nigeria to enforce
use of its local languages due to the multiplicity of diverse local
languages, despite the importance of Nigerian languages to the protection, preservation and promotion of cultures and to the enhance-
ment of human dignity, and the necessity of learning a major
language for purposes of promoting national unity and integration
as enshrined in the 1989 Constitution of the Republic of Nigeria. For
cultural identity and educational justification, the government has
settled on the use of immediate environment local language in the
first three years of schooling. Similar to the situation argued previously, in other African countries and in some countries in Asia the
English language serves as a powerful and prestigious tool in society.
It is time to reverse the order of power and prestige of the English language versus indigenous languages in schools. The ability to speak
indigenous languages should provide socio-economic benefits and
strengthen cultural identity. "Beyond pedagogic and psychological
reasons... language is inextricably linked to identity, ideology and
power" (Makalela, 2005, p. 163).

In Nigeria, teachers are poorly motivated, ill-trained, over-
worked, unevenly distributed and abysmally insufficient in numbers
(Emananjo, 2008). English remains one of the most poorly taught
subjects in the school system. Furthermore, most teachers of English,
including those with paper qualifications, are unqualified to teach

the English language, while most often redundant teachers are drafted to teach it. In Nigeria, as in Tanzania, it is presumed that anybody with a degree or a diploma is a potential English teacher (Babaci-Wilhite & Geo-JaJa, 2014a). The number of teachers required in 1988 for the three major Nigerian languages was 55,237, but only 6,383 or 11.6% of these were available. There is a great need to meet the demand for language teachers if cultural identity is a desired goal. The small and developing languages have no trained teachers, and for this reason the former National Language Center, now transformed into the current Language Development Center and placed under the Nigerian Educational Research and Development Council, in 1976 suggested that in addition to the three major languages – Hausa, Igbo and Yoruba – only the following nine of the remaining 387 or so indigenous languages in the country should be allowed to feature in the country's formal school system: Edo, Fulfulde, Ibibio, Idoma, Igala, Ijo, Kanuri, Nupe and Tiv (ADEA, 2001).

The Nigerian Policy of Education (2004) stated that:

> Government shall ensure that the medium of instruction is prin-
> cipally the mother or the language of the immediate commu-
> nity; and to this end will: (i) provide the orthography of many
> Nigerian languages, and (ii) produce textbooks in Nigerian lan-
> guages (Section 2:41e). At the primary level (e) the medium of
> instruction shall be the language of the environment for the
> first three years. During this period, English shall be taught as a
> subject. (f) From the fourth year, English shall progressively be
> used as a medium of instruction and the language of immediate
> environment and French shall be taught as subjects.
>
> (Section 4:19)

Furthermore, decree 16 of 1985 on Education (National Minimum Standards and Establishments of Institutions) gave legal backing and power of enforcement to the teaching of languages. All Nigerian languages can be used as MT or Language of Immediate Community. However, the pedagogical goal to organize initial literacy in 400 local languages can serve the needs of the educational process and become the medium for preserving the people's cultures. In theory they all qualify to be taught as school subjects under the Nigerian Policy of Education on language education in primary and junior secondary schools (ADEA, 2001). Clearly, because most of them have such small numbers of speakers, it would not appear to be practical to teach them as school subjects. However, the Nigerian Policy of Education also mentions that the language of the local environment shall be taught as a first language where it has orthography and literature. Where these do not exist, it shall be taught as a second language with emphasis on oral skills. At the secondary school level, competence in one major Nigerian language is a requirement for enrollment at the junior and senior certificate examinations. The national language policy contained in Articles 51, 55, 91 and 95 of the 1979 constitution states as follows: "The business of the National Assembly shall be conducted in English and in Hausa, Igbo and Yoruba when adequate arrangements have been made", while Article 97 of the constitution says that business in the State House of Assembly shall be conducted in English, but that the House may, in addition to English, conduct its business in one or more other languages spoken in the state. Soyoye (2010) asserts that these constitutional provisions, while not declaring English explicitly as the official language of Nigeria, give it the status of the priority language of governance at both the federal and state levels.

Afolayan argues that there are three primary functions for language in Nigerian education, making Nigerians capable of acquiring knowledge, skills and attitudes that will make Nigeria a highly developed nation by acknowledging "the importance of language in the educational process", making Nigerians capable of preserving and positively utilizing their cultures as "a means of preserving people's culture" and making Nigeria become a united nation "in the interest of national unity" (1990, pp. 5–6).

A comparative study of Zanzibar and Malaysia

In 2006, Zanzibar, which has a school system autonomous from that of the Tanzanian mainland, initiated a review of its educational strategy, which resulted in a reform of its curriculum. In addition to the change of LoI from Kiswahili to English in mathematics and science subjects from Grade 5, it introduced a new subject, Information, Communication and Technology (ICT), from Grade 5 in English, as well as Arabic from Grade 1. The new policy was supported by the Educational and Training World Bank project entitled "Zanzibar basic education of improvement project" implemented in

2007. The background for the new policy was that Zanzibar's educational program had achieved a number of successes but left unresolved problems such as quality education and English in secondary education.

In the period 2001–2010 the Ministry of Education, Culture and Sports established, within the three years of pre-secondary education, one year set aside for English called "OSC-Orientation to Secondary Class". The OSC is a class to prepare students for learning in English. Those who did well in English in primary school did not have to take this bridging year (Said, 2003). The background for the class is low English competence, which was affecting students' academic performance in secondary and higher education (ibid.).[2] The OSC was established by the Ministry of Education, Culture and Sports with the help of various international organizations such as the Aga Khan Foundation, ODA and Canadian International Development Agency (CIDA) and introduced in 1994. Said (2003) argues that it has not been successful in preparing the Zanzibari pupils to use English as a LoI. These problems were to be the targets of the policy reforms. Based on my research, there were other important reasons for the change, which have to do with a desire to reinforce local Islam-based culture and simultaneously to facilitate global integration through the increased use of English. However, based on an analysis of the early stages of implementation, my research shows that the change will in fact reduce learning and teaching capacity and that this constitutes a violation of children's rights in education.

The findings presented and discussed in my previous work (see Babaci-Wilhite, 2013) address the reasons behind the curriculum change and interrogate the relationship between choice of language, quality teaching and learning and children's rights in education. The choice of local LoI was found to be critical to the learning environment and to be inextricably linked to rights in education. In making the curriculum change, Zanzibari policy makers have been influenced by the still-powerful notion throughout Africa that learning in a "Western" language (English) will promote development and modernization. However, in the Zanzibari case, neither teacher preparation nor support materials in the form of books, teaching plans and teacher training have been robust enough to support the curriculum change. It was found that teachers are not fluent enough in English to master teaching in accordance with the new curriculum, nor have they been adequately prepared for the change of language.

Results indicate that teachers are skeptical about the capacity of the new curriculum to improve the quality of learning and the quality of English, as the poster (Figure 7.3) below shows:

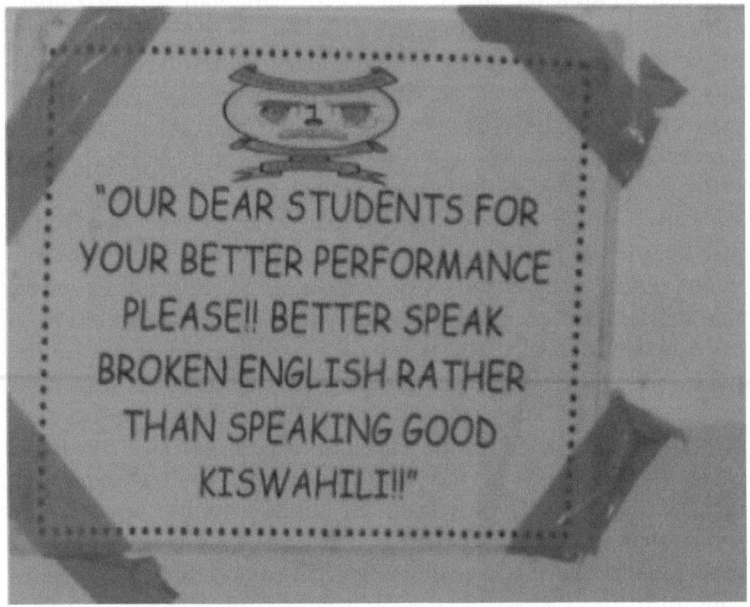

Figure 7.3   Photo of a poster in a classroom in Zanzibar[3]

Interviews with teachers revealed that they are losing their confidence, as the new curriculum does not reflect the reality on the ground and their preparation is not sufficiently robust. Their discomfort with the new curriculum and the increased use of English has affected teacher motivation and well-being, which is crucial for the success of the implementation. This was brought home to me during my interviews when the teachers had difficulties expressing themselves in English, which are reflected in the figures below when I compared the responses from my interviewees (Figures 7.4, 7.5, 7.6, 7.7, 7.8 and 7.9):

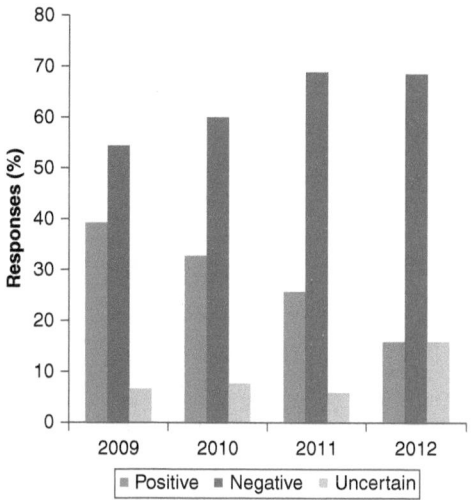

Figure 7.4   All interviewees' answers regarding the curriculum change

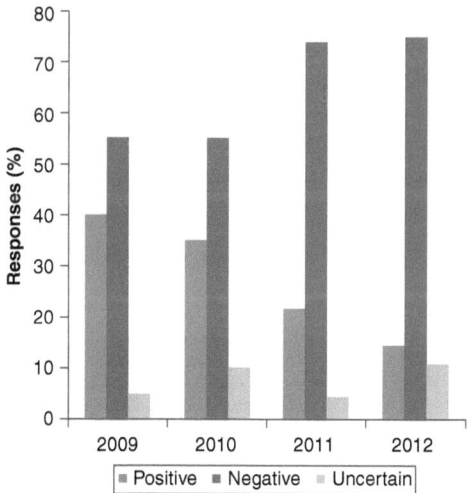

Figure 7.5   Urban and rural teachers' answers regarding the curriculum change

100

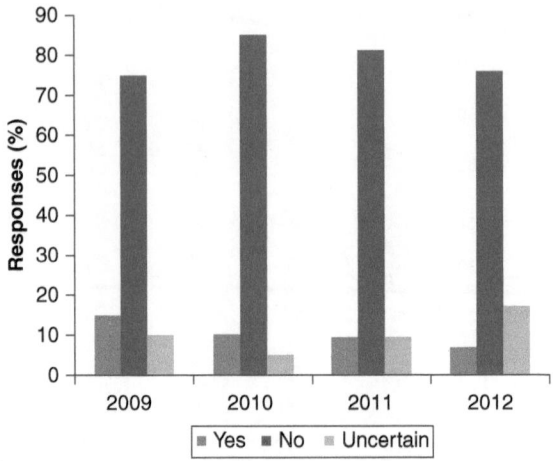

Figure 7.6 All interviewees' answers regarding the curriculum focus on quality of teaching and learning

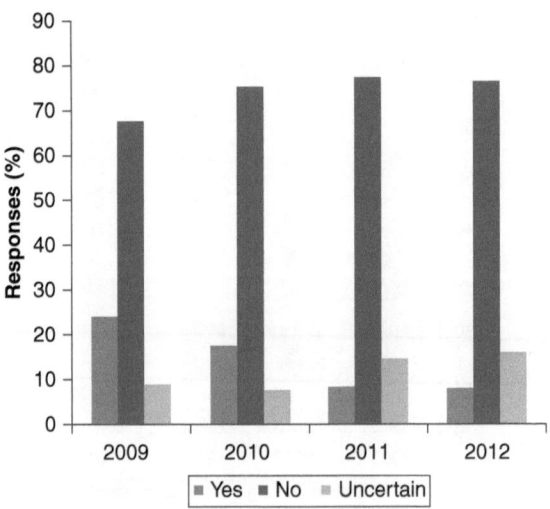

Figure 7.7 Urban and rural teachers' answers regarding the curriculum focus on quality of teaching and learning

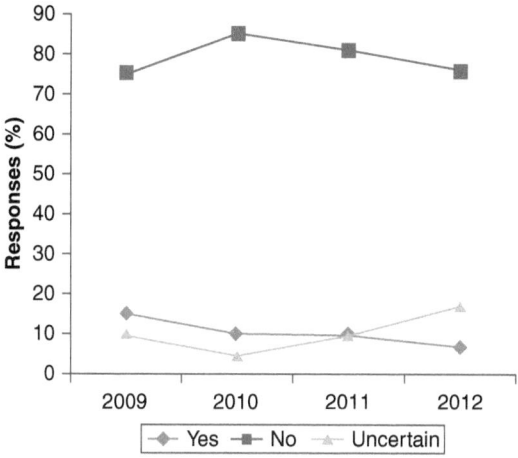

Figure 7.8   All interviewees' answers regarding the preparation for this change in teaching and learning through English as a LoI in science and mathematics

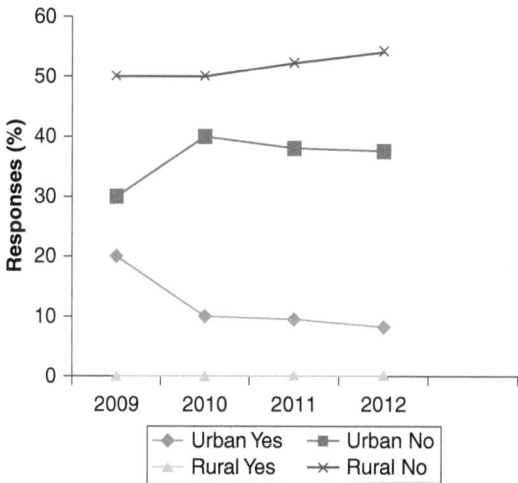

Figure 7.9   Urban and rural teachers' answers regarding the preparation for this change in teaching and learning through English as a LoI in science and mathematics

The graphs in the figures above highlight differences between all interviewees versus urban and rural teachers as well as urban and rural schools. The tabular analysis shows clearly that the teachers did not feel comfortable with the curriculum and they did not believe that the new curriculum would truly advance the quality of teaching and learning. The sequence of visits over time has been crucial to developing analytical insights. The implementation of the curriculum change is a work in progress and the methodological lens has benefitted from intervening at different points in the process.

One of the interviewees argued very strongly for teaching in Kiswahili and teaching English as a second language with the following argument:

> The problem remains that linguistic environment of Zanzibar favours the use of Kiswahili as a LoI. There is a problem of understanding what is a gap, what is a bridge, what is a language, we can teach people through Kiswahili and we can teach English as a foreign language. If we have enough preparation of language, people can communicate in English. We introduce people to foreign languages like French, Spanish and Italian, after two years people manage to communicate through French, through Spanish, through Italian. The solution is to teach English as a subject very effectively and efficiently.

He advises that the government favor Kiswahili because:

> In most cases people think very fast in its mother tongue, they think in Kiswahili, and then come other languages. There is a long process of imputing knowledge through the use of the second language. First of all a learner may conceptualize through Kiswahili then may translate to English and this is a long process of learning the acquisition of knowledge. At the university level, we need time at least for the preparation. To introduce Kiswahili in the secondary is not new, it was in the language commission in 1985, the forces in 1982 was suggesting that Kiswahili was in secondary school and 1985 should be introduce at the university. But the mission and the process was hijacked somewhere.

Given the deterioration of the learning environment attributable to the replacement of Kiswahili with English as LoI from Grade 5 in mathematics and science and the lack of adequate support for the change, my assessment is that the capacity for the curriculum to enable quality learning has been reduced. From the perspective of a Rights-Capability-Based Approach, this curriculum change violates Zanzibari children's rights in education.

Malaysia chose to reduce the role and status of English and select one autochthonous language, Bahasa Melayu, after its independence in 1957 (Gill, 2005). Furthermore, he states that English was relegated to being taught in schools as a second language and that in rural areas where there was almost no environmental exposure to the language; English was virtually a foreign language. Moreover, Gill (2005, p. 8) quotes Hassan (1988, p. 38), who explained that "The government set up a team of Malaysian and Indonesian language planners and academicians, including scientists who held a total of 6 joint meetings over a period of 16 years from 1972 to 1988."

Furthermore, the government of Malaysia, then called Malaya, chose Malayor Bahasa Melayu as its national language (ibid.). During that period of strong nationalism, the government did not feel the need to change the name of the language. Later, the racial tensions of the 1960s spurred the government to rename the national language as Bahasa Malaysia, the language of Malaysians. According to Asmah, "these two terms are used interchangeably – Bahasa Melayu to signify the language of the Malays and Bahasa Malaysia, to signify the language of Malaysians" (Asmah, 1992, p. 157).

Bahasa Malaysia was made the official language of government and the LoI at all levels. One of the motives behind this policy was to unite the linguistically diverse groups in Malaysia (Heng & Tan, 2006). However, the government did not attempt to control language use in the private sector, including business and industry, where increasing integration into the global economy contributed to a growing demand for English. In 2003, the government reversed its LoI policy, changing the language used in teaching mathematics and sciences in Malaysia to English (Prah & Brock-Utne, 2009). According to Cavanagh (2009), one of the drivers of the change were the demands from local governments, both in urban areas, and in rural areas of the country where the use of English is less common.

Six years later, in January 2010, the Malaysian government made a decision to reverse this policy (effective from 2012). The prime minister reported to the Parliament that students had not shown significant improvement in these subjects over the six years in which they were taught in English (Zalkapli, 2010). Another factor in the decision was growing protests from residents of rural areas. Representatives of local governments and schools argued that students from rural areas were disadvantaged in national tests because of their lack of familiarity with English.

Given that Zanzibar and Malaysia are moving in opposite directions on the choice of language in science education, a review of the theories, practices and politics behind these decisions promises to shed light on the broader debates on the consequences for learning, cultural identity and global integration. As Rwantabagu (2011, p. 460) argues, "Eurocentric in nature in the sense of ensuring the supremacy of English or French as the dominant channel of communication, knowledge acquisition and cultural production." He goes on to quote the statement by Arnove and Arnove (1998, p. 2) that "language policies are central to an understanding of how colonial powers attempted to use schools to assimilate, acculturate and control colonized populations". There is good reason to ask when African policy makers will revise their language policies according to contemporary reality and pay attention to developments in Malaysia and other Asian countries which fully rely on their own languages in education.

In Zanzibar as well as the rest of Tanzania, in the 1980s the government gave consideration to implementing Kiswahili as a LoI, but in the end did not follow through at all levels. It was argued that Kiswahili was not ready to be a LoI because there was a lack of books and terminology. The Chief Academic Officer at the National Kiswahili Council told me in an interview in Dar-es-Salaam that, "In the 1980s those arguments were ok, but now they are using the same argument even if everything is ready." Academics have tried to convince the government that Kiswahili is mature. Since the 1980s both book publishers and the National Kiswahili Council have engaged in the development of scientific terminology in Kiswahili. The Chief Academic Officer concluded, "We have enough dictionaries now, and we try to convince the government". In her opinion, it is wrong that the country does not use its language, Kiswahili, when the National Kiswahili Council has developed all the necessary terminology.

The government has stated that one barrier to using Kiswahili as LoI is the cost of translating the manuals from English to Kiswahili.

The government does not have the finances to translate all the secondary school manuals from English to Kiswahili. However, the PITRO II project has received funds to write books in Kiswahili for Forms 1 and 2 (Mulokozi et al., 2008), and all the manuals have been written by secondary school teachers.[4] The reasoning behind the project is that not only will those manuals help students to better understand the concepts in their own language and to think more quickly, but the manuals could also help in adult education. One academic argued:

> Yes, it is expensive, but not very expensive. It is just a matter of translation, a matter of translating the document from English to Kiswahili or Kiswahili to English. It is not expensive as such, but again it is the argument from the World Bank and other development partners to favor English.

This answer raises the question: why do the World Bank and other development partners favor English (Mazrui, 1997; Brock-Utne, 2007)? One of my interviewees concluded that:

> The change is made, but I still say that the knowledge cannot be acquired if the LoI is not understood. The teachers are not competent to teach in English therefore it is very difficult to impute knowledge. I think we have to conduct a research to find out the possible and visible LoI. But before doing it we have to educate society, community, and we have to educate them through giving concrete examples of the contents, which employ their language to impacts knowledge, this is very important. In fact there are many ways, first through radio programmes, through NGOs, mass media, through public talk and public lectures.

Gill (2005) describes a similar development in Malaysia. He explains that the development of terminology – about half a million new words had been developed by the mid-1980s – was considered one of the most significant achievements in language planning in the region, showing strong government support in modernizing the language in the post-independence period.

I found resistance to accepting these findings on language and learning in the Ministry of Education. The resistance is remarkable because of the consensus both within Tanzanian academia and abroad that the choice of a local LoI is important for good learning. In Tanzania, some of my academic informants proposed a kind of dual system. One of them argued:

> Once I went to a seminar and we were talking on Zanzibar issues and all of us are Zanzibari but most of the people addressed each other in English. We are Zanzibari, discussing the issues of Zanzibar for the benefits of Zanzibari but we addressed issues in English. It shows the problem that we are facing. People think that if we speak Kiswahili, we will be interpreted as ignorant. It is two different issues knowledge and language, although language is a mean of translating knowledge.

The same interviewee added:

> I was a product of that, we were prepared in English at the primary level it was not a problem, and this is my case. Why because it was a good preparation of both teachers and learners, there is not enough preparation of both learners and teachers. Teacher training college to me is not a right place to "cook" the teachers. There are many reasons including shortage of teaching and learning materials, shortage of qualified and motivated teachers that is why we need teachers who can provide a product inquired that is the problem that we are having.

According to the interviewee, the new curriculum is too packed for the children. To start learning two foreign languages in Grade 1 is a "burden"; a child should first master one language, then s/he can learn through and introduce respectively a foreign language. Research shows that the child should build first on what they have (Loona, 1996; Roy-Campbell & Qorro, 1997; Malekela, 2003; Vuzo, 2009) and then they can learn an unlimited number of languages in a later stage of their education. He argued:

> Not only that I favour Kiswahili but this is a viable mechanism to impact knowledge. In my view any project related to Education is capital intensive. It needs a lot of money but before investing using the amount of money we should conduct research. For this loan, we did not conduct research and that is why we got

a crossroad, rather we go this way or that way because of the stakeholder conditions.

He felt that the study behind the reform was predetermined, conducted in order to support the interests of the Ministry of Education and Vocational Training (MoEVT) and the donors. It was predetermined because the situation is contradictory to the results; the situation is opposite to the reality.

To draw out the threads of this comparison, Kiswahili and Bahasa Malaysia were chosen in Tanzania – both Zanzibar and mainland – and Malaysia respectively as the national language to unite the linguistically diverse groups in both countries after independence. This choice has contributed to the formation of a national identity incorporating these languages. There is evidence that having one common language helps in bridging the gap between people of different ethnic groups (David, 2007). Both Nyerere in Tanzania and the government of Malaysia were reflexive about these cultural identity issues in their choices. As Gill (2005) states, Malaysia focused like a number of other countries on the essential which he describes as "educational agendas of nation-building, national identity and unity". For this reason, the latest decision in Malaysia, which saw a complete reversal of the role of English, is very interesting in relation to the indecisiveness of the Tanzanian government when it comes to switching the LoI in secondary school to Kiswahili.

To summarize, the comparison of language-in-education policies in Tanzania, Nigeria, Zanzibar and Malaysia show that the governments of these countries have all been interested in maintaining cultural identity, national unity and social equality through the use of local languages and local curriculum in schools. It is perplexing that Zanzibar's recent decision seems to go against its own intention to preserve culture and produce quality education. Retaining local languages as a LoI will provide quality education. For the reasons I have outlined, children of all backgrounds will be able to perform better in school.

8
Experiences in Countries That Have Chosen English as the Language of Instruction

In this chapter I will analyze the successes and failures in selected Asian and African countries that have chosen to use English as the LoI at the upper levels of education. I will examine the linkages between LoI and cultural identity, economic development, work prospects and integration into the global economy. I will focus on two countries experiencing rapid economic development, India and South Africa, which are included among the BRICS countries (Brazil, Russia, India, China and South Africa), as well as two economically disadvantaged countries, Bangladesh and Rwanda. I will give some background on why the governments in these countries instituted English as a LoI and on the successes and failures of this policy.

Both India and South Africa have a multiplicity of languages, use local languages in lower levels of education and use English in secondary schools and universities, though in India there is an option in most states to use a local language as LoI at upper levels. Bangladesh and Rwanda have a single national language spoken by the majority of the people. Kinyarwanda is spoken by 99.4% of the population of Rwanda, while Bangla – also known as Bengali – is spoken by 98% of the population of Bangladesh. In spite of having their own national languages, English is taking over as a LoI in the secondary levels of most schools in both countries.

After exploring the reasons behind these choices of a foreign language as LoI, I will develop the association between language choices and human rights in education. The results of this chapter will have relevance for researchers examining the synergy between language and cultural identity, as well as for rights in education for quality education and inclusion.

The case of India

India is the second-most populous country in the world, with a population of over 1.21 billion, and it is the largest democracy in the world. The country is divided into 27 linguistically organized states. There are two national languages, Hindi and English. The language policy issues that have evolved since independence in 1947 have generated controversy regarding LoI at primary, secondary and university levels. Most Indian states have adapted the "three-language formula", which means that either a local language or one of the two national languages can be used as the LoI. The three-language formula emerged as a political consensus on languages in school education to accommodate at least three languages within the ten years of required schooling.

With regard to the number of languages taught, 90% of schools at the upper primary stage follow the three-language formula, and 85% of schools follow the formula at the secondary stage (NCERT, 2007). Forty-seven languages are used as LoI in schools and 41 languages are taught as a second language in schools (Srinivasa Rao, 2008). However, LoI in all colleges and universities is either a regional language or English.

There has been a general consensus that students opting to be educated in English take the local language as a subject, while students opting to be educated in Hindi or their local language study English as a foreign language from Grade 5. While public schools tend to use either Hindi or another local language as LoI, private schools usually prefer English.

Language policies in India today have to be seen in light of post-independence policies, in which Mahatma Gandhi played a major role. Gandhi believed that the use of an Indian language was essential to empowerment and maintenance of Indian identity. According to L. Chatterjee (1992), English was not considered desirable as an official LoI because it was seen as undemocratic and divisive. He wrote that it divided "the Indian people into two nations: the few who govern, and the many who are governed" (p. 300). Gandhi's pro-indigenous stances can be seen as part of his call for freedom and national unity, tied indivisibly to his views on LoI. He consistently maintained that a new, liberated India could only fully emerge if it completely embraced Indian culture and gave up being enslaved by all things British, including, of course, the crucial instrument of colonization – the English language. His primary goal was to create an environment where the poor could be empowered, working in local communities and building on indigenous roots. Gandhi advocated the use of indigenous knowledge in all realms of existence. Furthermore, he argued that the use of a foreign LoI made children imitators, unfit for original work and thought and unable to pass their learning on to family or society. He argued that "the foreign medium made our children practically foreigners in their own land. It is the greatest tragedy of the existing system. The foreign medium has prevented the growth of our indigenous languages" (Gandhi, 1954 translated by B. Kumarappa).

Today, some of the teachers that are teaching in a local language make explicit references to the relevance of Gandhi's ideologies. A school teacher interviewed by Vaidehi Ramanathan in 2004 referred to Gandhi when she made the claim that key terms in mathematics and science will not be developed in her state's language, Gujarat, if English is the LoI. Ramanathan interviewed another secondary school literature teacher who gave a good description of the impact of the LoI, arguing that her job is to give weight to indigenous ways of thinking, reasoning and believing, which through the use of a MT allows students to experience literature in a way that can be quite different from English. The way she sees it, the indigenous language empowers learners, while English does not. A general sense that emerges from statements such as these is that many teachers view their ways of teaching, learning and living as diametrically opposed to English language teaching and its general associations, the implication being that in India all things English "suppress", "disempower" and "devalue", in the words of Ramanathan.

Others contend that there is an "English craze" in India, reflected in English programs on television and state government policies to introduce English instruction in indigenous medium schools, as well as to encourage Westernized ways of dressing and speaking. Ramanathan (2004) quotes a Gujarati faculty member who said that his children are "crazy about English" because of their interest in British and American programs on cable television. This English craze is also reinforced by job announcements that require the applicant to be fluent in English. The students who study in an indigenous language recognize that they run the risk of being left out of the country's growing computer industry.

Ramanathan argues that Indians have a lot to gain from knowing English and that the world has a lot to gain from Indians knowing English. In his view, rather than worrying about whether or not English should be used, people should focus on providing more children with an education that permits them to learn and use English as a second language but also strongly emphasizes the use and understanding of their native languages. In the 1980/1981census, it was recorded that only *3% of India's population spoke English.* Twenty-five years later, the data from the 2005 India Human Development Survey estimated that *4% of the Indian population spoke English, an increase of only 1%.* Nonetheless, with India's massive population, that 4% puts India among the top four countries in the world ranked by number of English speakers. English confers many advantages on the people who speak it, which has allowed the language to retain its prominence despite the strong opposition to English that arises periodically in India. Private schools have mushroomed in recent years following the economic liberalization policy that began in the 1990s, and most of them have English as a LoI.

Vedpratap Vaidik (2004) analyzed the efforts to make English the LoI in Indian schools and concluded that it has rendered hundreds of thousands of children handicapped. He argued that only a very small percentage of the country's population has control over the country's resources. English has divided Indians from other Indians. Furthermore, Vaidik argued that without removing English as a LoI, we cannot eradicate poverty, since those who believe that English should be the LoI also believe that the world's knowledge is produced only in English, whereas in reality there are all kinds of domains in which progress has been made in a range of other languages. The result of India's LoI policies is a vast disparity in quality and access to education, a gap the government is trying to close. Despite this larger challenge, the Indian government has realized that preparing students for the global economy is important and must be a part of their reform. However, according to Ramanathan (2004), Indians should remember that the development of India today has been made on what "we know best: our languages, our culture, and our people". In my view, this should be a lesson for other multilingual countries.

The case of South Africa

South Africa is a multilingual society that has some unique linguistic problems that can be related to policies during the period of apartheid. Under the apartheid regime, South Africa's policy was to use African languages as LoI throughout primary education (up to Grade 6). There are approximately 25 languages spoken in South Africa, 11 of which have official status, including English and Afrikaans (Kamwangamalu, 2003). There have been significant cultural and linguistic tensions between Afrikaans and English. The conflict and rivalry between speakers of English and Afrikaans goes to the heart of the history of the original European settlers in 1652 that spoke Dutch, which eventually evolved into Afrikaans. But in 1822 the British gained control and proclaimed English as the language of the schools, churches and government. Afrikaners, who today constitute roughly 60% of the white population, dominated national politics after 1948. English, however, holds greater prestige in finance and education, and is the language of the more liberal newspapers as well as most black South African newspapers. In 1955, a policy of teaching in both English and Afrikaans on a 50–50 basis in the secondary schools was adopted.

In 1976, in conjunction with the protest of Soweto, black South Africans began referring to Afrikaans as the "language of the oppressor" because Afrikaans had been the language of the labor bureaus, the police and the prisons. This is part of the explanation for black South Africans' antipathy toward Afrikaans and their favoring of English as a LoI. As a result, the government has allowed individual school boards to choose the LoI, and 99% have chosen English. The group designated as Coloreds, a mixed-race group, has traditionally spoken Afrikaans, but many are now switching to English in sympathy with black South Africans.

Today, the Language in Education Policy of South Africa states that multilingualism is an essential goal of the South African education system, with the choices of LoI drawn from one of the 11 official languages. Nevertheless, Kamwangamalu refers to MacFarlane (2012) who clarified that contrary to the policy, most schools currently adopt an early-exit model with a transition to English as a LoI in Grade 4 despite the fact that only 9% of the population speaks English (Kamwangamalu, 2003). Although there is a strong motivation for "black" Africans to learn English, there is also an interest in African languages, with newspapers and literature in languages related to as Bantu. About 70% of South Africans speak one of the Bantu languages. There is an almost perfect correlation between race and language: those who speak a Bantu language as a MT are black South Africans, and there are very few black South Africans who do not speak a Bantu language. Schools and universities generally use either English or Afrikaans as the LoI. Because the MTs have been imposed by the government in order to distance black South Africans from the well-established white power structure, their use has been resisted. The government's divide-and-conquer approach to language policy is allied to the overall degrading system of laws that keep black South Africans in permanent poverty. Language and politics are very much intertwined in South Africa, as they are everywhere.

According to Ayo Bamgbose (2003), owing to agitation against this policy, particularly its accompanying imposition of a dual medium of Afrikaans and English at post-primary level, the MT medium was whittled down progressively from eight to six, four, and, eventually, zero in South Africa. The position today is that African languages are not generally used as LoI in secondary schools or even in primary schools, except in pilot projects such as mixed English and Xhosa classes in the Department of Education and Training schools in the Western Cape, which still carry on teaching and learning in isiXhosa, Setswana and Sesotho. This situation has contributed to the creation of two social classes, the elite and the others. The elite can speak and write the official language, but are usually illiterate in the native languages. Therefore, they cannot communicate effectively with other South Africans. The issue has been also taken up by Zubeida Desai (2004), one of the leaders of the Language of Instruction in Tanzania and South Africa project (LOITASA), who points out that:

> In South Africa too, MT education is seen as a given for English-speaking, and to a lesser extent, Afrikaans-speaking learners. It is taken for granted that these learners will learn best through their primary languages, however, and when it comes to speakers of

African languages, the debate rages furiously. Why is this "right" so wrong for the majority of learners in African countries such as South Africa?

In order to maintain cultural heritage and at the same time promote communication among members of the same ethnic community, with other communities and at the international level, it is important in these contexts that students should master their native languages as well as official languages both orally and in writing. This can be achieved only by using local languages as LoI at upper secondary education.

The case of Bangladesh

Bangladesh is a nation of 133.4 million people and one of the most densely populated countries in the world. Language played a crucial role in the country's struggle for independence from Pakistan in 1971. In 1972, Bangladesh adopted its constitution in which it declared Bangla – also known as Bengali – the "state" language and established Bangla as the LoI in primary and secondary schools (Constitution of Bangladesh, 1972, p. 3). The background report claimed that Bangla has many advantages as LoI, particularly its value in developing students' "natural intelligence", original thinking and imagination (Ministry of Education, 1974, p. 14).

Bangla is spoken by 98% of the people of Bangladesh and is also widely spoken in India, where it is the dominant language in the state of West Bengal, including the city of Calcutta with 58.5 million speakers (Agnihotri, 2001, p. 190). In 1999, in recognition of the Bangladesh language movement, UNESCO adopted a resolution declaring February 21st "International Mother Language Day", the day in 1952 that Bengalis sacrificed their lives in the struggle for their national language. A Bangla Academy was set up to promote the development of Bengali language and culture.

In 1998, the Ministry of Education signed an agreement with the US government for the promotion of English in secondary education. Every year 200 American volunteers travel to Bangladesh to teach English in secondary schools. In addition, English teacher training specialists from abroad come to Bangladesh to teach second-language teachers English, using British Council funding. In 2002, the president of Bangladesh stated that he would put an emphasis on teaching the English language with a view to promoting employment and encouraging transfer of technology. According to UNESCO, in 2007, many of the secondary schools in Bangladesh became private, adopting one of three educational policies: Bangla-medium schools, where English is taught as a compulsory subject, whereas most classes and informal interaction take place in Bangla; English-medium schools, where Bangla is used for much of the informal social interaction but English is used for subject-matter instruction; and Madrasah religious schools, where Bangla and Arabic are used as LoI and English is taught as a subject.

In the private English-medium schools, all courses are taught in English using books that are produced in the UK, except for courses in Bangla. The core part of this sector is the small number of leading English-medium schools charging high fees, which are much better resourced than public schools. Essentially these schools provide a globalized curriculum, imported from the UK, for the preparation of the social elite. English-medium schools are mainly private and thus reserved for the wealthy class.

The elitist role of English in Bangladesh is now being enhanced by World Bank strategies designed to promote the private school sector. The World Bank and the Asian Development Bank argue that the most cost-effective means of expanding educational provision is a greater investment in private schools, private universities and non-formal adult learning. Bangladesh is being exposed to English as never before. The entrance of Bangladesh into the garment industry and global trade has created an increasing awareness of the need for English communication skills. The phenomenal growth of the Information Technology industry in Bangladesh has also made people aware of the importance of English as a language of communication. English has become essential for business and economic purposes.

Around this elite core of English-medium schools are less exclusive schools charging a lower level of tuition and providing on the whole a lower quality English-medium instruction (Chowdhury, 2001). These schools, growing in number, are designed to capture the market among non-elite families for English language education. Teachers at these schools are neither required to have completed English-medium schooling, nor to have taken English at tertiary

level. The students do not speak English as well as their counterparts at the elite schools, though they learn to read and write fairly
well (Chowdhury, 2001). In the later stages of schooling, from Grade
8 to A level, these students use the same texts as their elite school
counterparts. A small number of low-cost Bengali-medium schools
use the English language as a medium of instruction. In both cases,
the schools follow a Bangla curriculum translated into English, rather
than a British sanctioned curriculum.

The private universities also provide English-medium instruction.
In addition, the Cadet Colleges (military schools) decided in 2002
to use English-medium instruction. At the public universities, apart
from the English departments, students have the option of answer-
ing examination questions in either Bangla or English. The elite
English-medium schools in Bangladesh offer the most prestigious
educational commodity within the private sector. Significantly, the
commodity on sale in this subsidized educational market is not only
English language but also English culture. The global utility of private
English-medium education is one of its main attractions. It is note-
worthy that recently some embassies and high commissions, have
founded their own schools in Bangladesh in response to the demand
of "standard" English-medium schooling. Examples are the American
School and the Australian High Commission School. The spread of
English also facilitates those NGOs that use programs of aid, credit
or business activities through microfinance (Chowdhury, 2001) to
control key social and governmental policies.

The case of Rwanda

Rwanda has a population estimated at 11,689,696. Language policy
in Rwanda has revolved around three languages: Kinyarwanda, the
indigenous language of Rwandans spoken by 99.4% of the population; French, spoken by 8% of the population; and English, spoken
by 4% of the population. Rwanda was a colony of Belgium and maintained French as a LoI, even though it had its own widely spoken
language, Kinyarwanda (Rwanda, 2005, p. 38). Kinyarwanda uni-
fies the population because, unlike most other African countries,
Rwanda has only this one indigenous language. Despite its uni-
fying characteristics, Kinyarwanda has been neglected as a LoI in
schools.

French was the official language in Rwanda under Belgian rule from 1890 to 1962, but in the aftermath of the 1994 genocide the Rwandan government has worked strenuously to construct a new national image. The shift in language policy from French to English is part of this ambitious project, with expected profound effects on the political and economic landscape of the country.

According to Izabela Steflja (2012) there are several reasons for the Rwandan government's decision to transition from French to English as the country's main official language. One of the reasons for this shift is the government's desire to join the East African Community and engage in economic relations with other member states neighbors and South Africa. Rwanda relies on trade with Uganda, Kenya and Tanzania and since 1994 Rwanda has increased its economic ties with the UK and the USA. Rwandan officials emphasize that their eagerness to switch languages in the education sector from French to English is in response to its perceived economic advantages. According to Rachel Hayman (2005) the UK and the World Bank have been the most influential development partners in terms of education policy making. The UK is now the largest bilateral donor to Rwanda, and the largest education sector donor.

From 1996 to 2008 the language policy required children to begin their study in English from the first grade and take their secondary school entrance exams in Grade 6 in English (Samuelson and Freedman, 2010, pp. 194–195). As of 2004–2008, 77% of the population was literate in French, which is a relatively high percentage; however, only 31% of students continue into secondary schooling. In 2008, about 40% of the teacher population in Rwanda had less than five years of experience in teaching English. The percentage of teachers qualified to teach English in secondary school is only 36% for lower secondary and 33% for upper secondary.

Adopting English for the purpose of economic and social betterment and as a "language of progress" put English-speaking groups at an advantage in socio-economic relations, and non-English-speaking groups at a disadvantage. According to Samuelson and Freedman (2001, p. 203), "the Rwandan population has a positive attitude towards the use of English language: they perceive English as a valuable commodity". However, it is important to keep Rwandan realities in perspective when examining policies. The government has taken ambitious steps to change how Rwandans perceive their identity. In an effort to "eradicate genocide ideology", the government is hoping to eliminate affiliations based on ethnicity and create a single national identity. "The way to heal the divide and heal Rwanda is to promote Rwandan identity above all other identities", meaning "Rwanda first, Hutu and Tutsi later" (Whitelaw, 2007); however, one can question why English is being promoted and not Kinyarwanda. English is the language of Rwanda's elite and the language of the new leadership, as well as of other Tutsis raised in exile in Anglophone countries. This includes the Tutsi members who organized the invasion from Uganda which stopped the genocide, as well as many Tutsis who were in exile in Uganda, Congo, Burundi, Tanzania, Kenya and South Africa and returned to Rwanda after 1994, most of whom are Anglophone.

The government of Rwanda agreed on a trilingual education policy – including Kinyarwanda, French and English – to secure greater equity between groups who favored one or the other language. Michele Schweisfurth (2006, p. 703) notes, however, that development partners, on the other hand, expressed concern for the potential impact of a trilingual policy, claiming that learners struggling in one language may be further handicapped by having to cope with three and that quality in education, as a dimension of EFA, may suffer. Schweisfurth (2006) contends that this policy favors "the children of the educated elite who can offer expensive private schooling, extra tutoring, assistance at home and extra resources, the masses of African children are not". I agree with Brock-Utne's (2012) argument that a trilingual policy might have been good for Rwanda provided that Kinyarwanda had been the LoI and French and English taught as foreign languages.

Drawing together the experiences of these four countries, it is time to recognize the wealth of African knowledge and to promote its languages in education at all levels, including secondary and university. This would make a significant contribution to African development on its own terms and for the benefit of the majority of Africans.

The use of English at upper levels of education in African and Asian countries serves to increase socio-economic inequality. In many African countries, English is the language of secondary and tertiary education for the vast majority. While English has produced an elite, we have to find ways of bridging the gaps between the rich and the poor and empowering people in disadvantaged socio-economic groups so that they too can participate in their country's development. A "bridging" policy to enable children to make this move, from the use of the MT in the early years of their education to the use of a national local language at a later stage of schooling, could be the answer. As Kwaa-Prah (2009) notes, it is not an exaggeration to say that African nations which have many local languages but which make use of the former colonial language as their national language look with envy at countries such as Korea, Japan or China which possess their own well-developed national languages.

The approach of teaching other subjects through English in order to improve the quality of English carries many risks. Moreover, it is an approach that is resource intensive, requiring, for example, the employment of large numbers of foreign teachers or the retraining of even larger numbers of local teachers. Pennycook (1995) notes that English acts as a gatekeeper to positions of wealth and prestige both within and between nations and is the language through which much of the unequal distribution of wealth, resources and knowledge operates.

The World Bank sees education as a market, a notion that rests on private investment in global English as the source of individual advantage, an investment pursued without regard for the development of the countries that receive its support. One problematic result is that global English and international education are seen as a source of quality in education. Public schooling and local university education are relegated to lower status, unable to maintain the support of the elite or attract sufficient government investment. It is vital that national policy secures more effective control over educational development.

Several important questions remain unanswered: What does the English language have to do with development in non-English-speaking countries? Why does one need to adopt someone else's language/identity in order to develop? Is it not possible to develop within one's own language and identity? It is high time for the non-English-speaking developing countries to ask questions about whose political, economic and cultural interests are being served through the World Bank's language promotion and what measures can be taken to balance the situation.

A language policy is not an end in itself. The rationale for it must be what it can contribute to the overall cultural and socio-economic development of a country. In this connection, it is not enough to place emphasis on Western innovation, knowledge and language at the upper levels of education, to the detriment of local languages through which most of the population can participate and make any meaningful contribution to national development.

9
Science Literacy and Mathematics as a Human Right

In this chapter I will address why many educational practitioners and policy makers continue to ignore local languages and culture as central ingredients in science literacy and the consequences of this policy for learning. As discussed in Chapter 7, past projects in Africa and especially in Nigeria, as well as case studies in some Asian countries, have shown how literacy and science can be better achieved in a local language, yet Zanzibar recently changed its LoI from a local language to English in science and mathematics.

The role of language in science literacy

There has been a tendency in academic approaches to science literacy to regard literacy as an end unto itself, ignoring structures that undercut disciplinary learning, comprehension, critical literacy and strategic reading (Cervetti et al., 2005). Michel Foucault (1988) claims that belief systems gain momentum and hence power as more people come to accept as common knowledge the particular views associated with that belief system. This insight is highly relevant in science and language politics. In order to address claims about the importance of English and other global languages in education, it is important to look at the ways political and economic globalizing forces manifest themselves in particular cultural settings. Freire's (1970) pedagogical theory on learning and the role of schooling in education is important in this context. He views learning as a critical process consisting of reflecting, unlearning conventional truths and relearning, a process in which the valuation of local knowledge is important. This point is also echoed in Chambers' (1997) theory on "Whose knowledge counts?", as well as Sleeter's (2001) related work on changing definitions and interpretations of "knowledge" belief and culture, and their implications for pedagogy. Bishop and Glynn (1999) argue along the same lines that culture and the range of socially constituted traditions for sense-making are central to learning. Since culture, sense making and language are intimately related, there are strong arguments for using a local language in learning.

An important consideration in understanding why local knowledge and local language have been "subjugated", in the words of Foucault, to Western theory and Western science is related to Western-based conceptualizations of "development", seen variously as modernization, or even as emulation, as well as discriminatory concepts from Western international relations theories such as "divisible sovereignty" or "semi-sovereignty" (Bull & Watson, 1995). These global discourses on meaning-making, ways of agreeing, thinking and conversing are brought into classrooms and contextualized in the LoI, as are the full range of ideas about political behavior, international relations, and diplomatic thinking about language, politics and power. Bishop and Glynn's (1999, p. 148) model implies "that it is essential that communities acknowledge their own diversity as they reflect on and develop their knowledge-*of*-practice".

David Pearson's approach to literacy has relevance for the language debate in Africa. His approach aims to develop the student's potential to use the information s/he gains from reading and apply it to new situations, problems or projects (Pearson, 2007; Cervetti et al., 2012). Pearson's approach emphasizes the role of language and literacy in supporting disciplinary learning, which can be achieved by using literacy skills to think critically and flexibly across many domains of knowledge and inquiry. In line with Pearson, Jacqueline Barber (2005) argues that "Inquiry is curiosity-driven... It involves reading books to find out what others have learned... Inquiry requires the use of critical and logical thinking... Good readers inquire information gathered from text." Pearson and Barber developed a model of teaching science literacy entitled "Seeds of Science/Roots of Reading" which acknowledges that knowledge and wider vocabulary are a consequence as well as a cause of reading comprehension. The aim of the Seeds of Science/Roots of Reading model is to make sense of the physical world through first- and second-hand experiences while addressing foundational dimensions of literacy. Therefore learning through local languages will improve literacy as well as the learning of the science subject.

These important issues of improving science learning and enabling rights in education can be addressed by applying the Seeds of Science/Roots of Reading model in education curricula. This will positively impact on the problems of quality of teaching and learning science associated with poorly prepared teachers and inadequate teaching aids and facilities (Babaci-Wilhite, 2014). A problem in science teaching is that the knowledge that forms the basis for school curricula is decontextualized. The educated person in one who has mastered sets of facts, propositions, models and cognitive skills that are fundamentally separate from the context in which they learn. Stanton Wortham and Kara Jackson (2012) argue that educational approaches differ in their methods for improving student knowledge, ranging from directives to more discovery-oriented pedagogies, but mainstream approaches all assume that student skills and representations are the target of educational interventions.

Knowledge is also typically viewed as relatively stable. In mainstream approaches to education, schooling often involves the transmission of isolated, portable bodies of knowledge and science. Schools make sense as institutions only because stable knowledge and reasoning procedures can allegedly transfer and have value in other contexts where students will use the knowledge they learned in school. Because the context is not integral to the knowledge or skill, the isolated bodies of scientific knowledge often hold little meaning for anyone other than the members of the community who generated that knowledge. The problems students solve in school are thus problems of the disciplinary communities from which the knowledge originated. This often makes schooled knowledge and skills less useful outside of schools. Moreover, given the de-contextualized, insular nature of the scientific knowledge being passed on, there is generally little opportunity for students to question the claims on which the knowledge is based (Babaci-Wilhite, 2014). Pearson and Hiebert (2013) argue that in the effort to promote understanding, existing background knowledge matters. They refer to Goldenberg (2008), who claims that:

> Beginning reading instruction is guided by neither a theory nor a goal of knowledge development. In fact, just the opposite: children are presented with texts – mostly narrative – chosen

to reflect their existing background knowledge, the assumption being that they can use that knowledge to comprehend familiar content.

From this perspective, engagement with local language and local knowledge are necessary to facilitate the teaching and learning process. This reflects the current situation in Nigeria as argued by Osungbemiro et al. (2013), that teaching science and technology in a local language requires that a large body of scientific literature be developed in that language using the appropriate scientific words, terms and phrases. Furthermore, they argue that:

> If everyone could speak the two most common used languages in their community then communication would be much better. The Federal Government of Nigeria should encourage the use of native and foreign languages as media of instruction by placing more premium on graduates who are versed in the effective use of an additional native language to the official *lingua franca*.
>
> (2013, p. 5)[1]

Recent research has been investigating the use of Yoruba in four schools in Ondo West Local Government Area. The team researchers, Osungbemiro et al. (2013), argue that "Teaching science and technology in the MT for example Yoruba requires that a large body of scientific literature be developed in such a mother tongue using the appropriate scientific words, terms and phrases." Furthermore, they claim that teaching biology involves more than translating concepts, terms and principles such as "pectoral girdle" or "odontoid process" to Yoruba. Equivalent concepts need to be developed from within the Nigerian cultural context.

> Teaching biology in Yoruba language provides a forum for enhancing a deeper understanding and knowledge of the topic. More importantly is the fact that interest of the student is aroused when some terms like cranium, scapula, cervical, thoraxic etc. hitherto expressed in English are now translated with their functions into mother tongue and this leads to more concentration on the part of students. Many scientific concepts, terms and principles which

do not have equivalents in Yoruba language can be absorbed through transliteration in spellings such words using Yoruba alphabets and of course, incorporating the total nature of the language.

(2013, p. 5)

Alternative assumptions about the value of local knowledge and local languages in the teaching and learning of science subjects in several aspects will improve the quality of learning. Teaching science in a local language will also improve literacy.

According to Osungbemiro et al. (2013), the use of Yoruba as LoI contributed to higher achievement among the experimental group. They conclude that the difficulty of teaching and learning science is due to

> its being too wide in nature was taught and learned with rela-
> tive ease using Yoruba language as medium of instruction. If we
> are interested in inculcating the spirit of science in our students,
> if we do not want to be left behind in the global world for sci-
> entific and technological development, then it is imperative for
> all concerned – science teachers, science educators, educational
> administrators, curriculum developers, authors and policy mak-
> ers – to join forces in order to teach and learn science in a language
> that gives minimum difficulty to the students (Olarewaju, 1991)
> thereby making the desired results a reality.
>
> (Osungbemiro et al., 2013, p. 8)

The authors of the study conclude: "Despite the government's commitment to education, the quality of education in our schools has been declining tremendously, thereby giving successive governments serious concern." Language is crucial to the learning process inside and outside of schools.

In most African countries, English language is introduced at an early age in order to ensure that the standard of written and spoken English of students is relatively high. However, Okonkwo (1983) reviews reports showing that pass rates and levels achieved by students in examinations, particularly in English language, science, technology and mathematics, at the General Certificate of Education Ordinary "O" Level conducted by the West African Examinations

Council. An analysis of available results from 1964 to 1972 reveals that the percentage of candidates who gained distinction or credit in English fluctuated between 19% and 24% while more than 30% of the candidates failed in the English language examination each year according to Aboderin (Olarenwaju, 2008). A similar conclusion was reached by the Nigeria Examination Sub-committee, which stated that performance reached its lowest ebb in 1985. The rate of distinction and credit pass fell to about 16% in 1979, while more than 50% failed. Available reports for the period from 1983 to 1985 also show that performance ranged from 5.14% to 14.49% at distinction/credit level, while failure rates ranged from 59.08% to 82.49% (Okanlawon, 1987). Language influences the thought process of the learner and his/her understanding of his/her environment (Olarenwaju, 2008). Consequently, in the learning of science, deliberate effort should be made to enable students to learn science in their MT.

The use of English and its impact in science and mathematics

The curriculum change in Zanzibar has, among other changes, replaced the current LoI, Kiswahili, with English in the subjects of mathematics and science from Grade 5 despite the fact that Zanzibar is the birthplace of Kiswahili,[2] which is the *lingua franca* of Tanzania and East Africa, and the MT of Zanzibari.

As mentioned in Chapter 5, Zanzibar began the implementation of the curriculum change in 2010. Important objectives of the reform are equal access, equitable treatment, improved quality, as well as gender parity in basic education. In 2002 enrollment in schools was very low, reaching only 65.5% for basic education and only about 14% for secondary education (ZEDP, 2006, p. 18). Another aim of the policy was to increase enrollment to 90% by 2012. Gender parity was also important. Such parity has been achieved in primary education but not in secondary or in post-secondary, where the level of male enrollment has been much higher than that of females (ZEDP, 2006, p. 9). The implementation started in 2010 for Grade 1, and continued until 2015 for the other grades. The primary school curriculum has been revised and reduced from seven to six years. One year was given to train the teachers for each grade until Grade 6, which covers all the primary years from January 2011 for Grade 2, from January 2012 for Grade 3, from January 2013 for Grade 4, from January 2014 for Grade 5 and from January 2015 for Grade 6.

The content of the reform was developed by a broad range of policy and educational experts, including the director for policy and planning, the director for curriculum, the directors for primary and secondary education at the MoEVT, as well as tutors from the Education College and representatives of grassroots organizations such as parent–teacher committees. Members of organizations outside Zanzibar were also involved, including World Bank staff, experts from Ireland and consultants from many parts of the world. According to local academicians whom I interviewed, the government's assessment of past problems and future solutions was based on incomplete information, giving most of its attention to the opinions of parents and too little to the assessments of teachers and students.

In this section I critically analyze these changes, how policy decisions were made and their consequences for quality education in science and mathematics. An important aspect of the analysis has been to examine whether the curriculum changes proposed in Zanzibar truly advance the quality of teaching and learning, and whether the changes will contribute to equal access to a quality education. My intention was to answer the following question: Will Zanzibar succeed in avoiding problems revealed in studies in other African countries which concluded that the curricula do not reflect local thinking in teaching and learning? I will also critically analyze the multiple "drivers" of education and development, which are increasingly influenced by global actors (Tandon, 2008). These drivers are an important factor in the ways political and economic globalizing forces manifest themselves in particular cultural settings. To understand how these changes came about, I focus on the importance of which knowledge – local or global – should be taught in schools, and on the power that various knowledge regimes have to influence local curricula (Nyerere, 1967; Chambers, 1997; Babaci-Wilhite, 2013b). Learning in a local language brings knowledge that is often more accurate and sophisticated than "Western" "scientific" knowledge. I will also draw on research relating LoI as a right in education (Babaci-Wilhite et al., 2012).

The OSC-Orientation to secondary Class did not help the learning in science and mathematics in English, the expert assessment suggested that English should be the LoI from Grade 5 in primary school in the subjects of mathematics, science, geography and ICT (MoEVT, 2007).[3] According to the policy assessment, language proficiency in Kiswahili and English are both very low, which obviously affects quality learning. The policy recommends that Kiswahili and English be taught as subjects as early as possible. All teachers should be competent and fluent in both Kiswahili and English. Language training should be strengthened in primary school using electronic media (ZEDP, 2006, p. 8).

The new policy focuses on teacher training with the purpose of assisting teachers in adapting to the new curriculum, and on teaching mathematics and science in English. The policy (ZEDP, 2006, p. 21) states that all teachers will undergo a 20-day training period in which they will all be trained in English competence using English as LoI in mathematics, science, geography, ICT and English; competence of language teaching in both Kiswahili and English from Grade 1 to Grade 4 will be upgraded. My observation of classroom activities and teacher–student interaction in the schools was crucial for my assessment of the achievement of quality learning in the classroom context. I gave particular attention to which language was used in various classroom situations and how this affected student interest, participation and comprehension. The observation enabled me to understand more clearly that the language of interaction was not English. In most situations, teachers refused to answer my questions in English. They clearly preferred to express themselves in Kiswahili, and this included the English teachers.

I interviewed the head teacher of each of four schools and 45 teachers of the subjects English, mathematics, science and Kiswahili. I conducted return interviews with some of them. Interviews were set up with each of the schools' head teachers and with a sample of 16 teachers: four from each of the four schools, in four different subjects, Kiswahili, English, mathematics and science. The head teachers organized several group interviews with teachers from different levels. I interviewed Grade 1 teachers in 2010, Grade 2 teachers in 2011 and Grade 3 teachers in 2012. Most teachers pointed to the lack of support in the form of books and other support materials, which they regarded as crucial to quality teaching

and learning in accordance with the new curriculum timetable
(Table 9.1):

Table 9.1   The new curriculum timetable in Zanzibar

Subjects	Standard 1–4 Before 2010	Standard 1–4 After 2010	Standard 5–6 Before 2010	Standard 5–6 After 2010	Standard 7
Kiswahili	10	8	6+ LoI	4	6
English	7	8	7	5+ LoI	6
Mathematics	10	8	8	6 (to be taught in English)	8
Environment Studies	6	0	0	0	0
Science	0	4	4	4 (to be taught in English)	4
Religion	4	3	3	2	3
Arabic	0	3	4	2	4
Sport	3	2	0	2	0
Civics/History/ Geography	0	4	6	6	6
Vocational	–	–	3	6	3
ICT	–	–	0	2 (to be taught in English)	0
Sum of periods per week	40	40	40	40	40

I have selected some answers from my interviewees regarding the
use of English and its impacts in mathematics and science. The
headmasters mentioned that:

> The programme for primary will start with teaching the English
> teachers from Grade one to Grade four, which is fine, but science

and mathematics is a problem because they are not going to use English as a LoI until 2015.

The interviewee at Nkrumah Teacher Training College stated that:

> We will start the in-service training, we will have a lot of microteaching, they will have small assignments to write and after 2 years we will have an assessment to show us how the teachers learned. We will start with primary because of the change of curriculum and then we will move to secondary, because before we had a programme for secondary school teachers and in this programme we were updating science and mathematics but at the same time English and it were successful.

One of the teachers said that "A lot of the teachers cannot speak English, we have tried to design and we discuss with our trainers many approaches and many effective ways with the Nkrumah Teacher Training College." There is another program that mainly focuses on sciences and mathematics. On this subject one of the trainers stated that:

> We do not have qualified teacher but in addition to that it will be another challenge when we will ask them to teach in English, they will come across many terminologies which are in English and they use to teach those terminologies in Kiswahili therefore we need to familiarize them in those terminologies, teach them to be able to communicate them that is another challenge.

The teacher advisors at the Nkrumah Teacher Training College have many tasks, because the English language is poor for both primary and secondary school teachers. She stated that:

> We can teach them simultaneously, content, methodology and language. If you teach them only content it is wastage because they lack the language required to perform subject. The one they find difficult was the science, mathematics, physics, and chemistry. The teachers find those four modules so difficult that

sometimes they skip the topics. Therefore we teach those methodologies and other approaches.

The major changes in the curriculum have to do with the way sciences are taught, she said:

> Before they had science but they had one lesson named Mazingira (Environment). They were learning politics, science, and geography but now it is several things. Now we are separated, they have different subjects. Another change will be English language from Grade 5 because before all the lessons were taught in Kiswahili except English language but now almost all the lessons will be taught in English, the change is good. When I was a child science and geography were taught in English, so we are going back to the old system. The main problem of today is English because the mainland is sending us the exam in English and our students who know only Kiswahili cannot pass the exam. They are good in Science but they do not know English that is why they fail the exams.

However it was clear from the interview that she was confused about the difference between understanding English and using English as a tool to learn science. Still, her analysis of the problems related to having the exam in English is correct. If the student cannot express her or his knowledge in English, s/he will fail the exam – not due to lack of competence, but because s/he does not know the language. The evaluation will then be incorrect. The other teacher added:

> Before the children could learn in English to improve their English. Mix both language could be used for them to understand. If they could use English in nursery that will help. The best way to teach mathematics is to use local context. For example to use leaves, lemon, sun flowers to learn algebra ex. Lemons begin with a L and if you have 5 lemons with 5 L we can say 5 L – 3 L = 2 L.

He added that to teach geometry they can use the shape of school gardens, while to teach ratios they can use food measurements, for example, to know how much food they need to cook for a given number of people or for building a house with correct measurements. He did not know a word of English. Another teacher who did not speak English, for whom the school principal served as interpreter from Kiswahili, said:

> I do not understand the new curriculum, no training, no books, and no aids. I think in the new curriculum, they want to change the LoI and few subjects... they added new topics like "air" I do not understand it.

According to the interviewees, the policy changes are not based on solid research. Rather, they are the result of a political rather than an academic decision. One academic argued that:

> One may have knowledge through any language, but people do not understand. If people do not speak English, they will say they do not have knowledge, although in other circumstances he is knowledgeable with let see indigenous thinking, knowledgeable with science, with other skills but people will think that you are illiterate because you do not write English. Even the people in government sometimes are confused because they are incompetent in this issue, they will just follow the general opinion of the people. But as a government, they need a very scientific research to reach a conclusion, but again it depends on your data, the process of collecting your data, this is very important. If your sample is wrong, automatically your data will be wrong. But frankly speaking the government is incompetent on this point.

The academics emphasized the importance of forcing politicians to acknowledge the experiences of other countries that employ their MT as LoI, such as China, Japan and many other countries, among them all European countries (Rubagumya, 2003; Prah & Brock-Utne, 2009). All of these have performed well in science learning.

When it comes to content, many of the academicians agreed that most of it is good because it is localized. One of the interviewees said:

I have participated in the content, and we localized everything even the methodology. That is what we call localization of knowledge. If you talk about river, you can start with the small rivers around. At least the students will get an idea about the thing. If you talk about mountains, you do not have to start with Kilimanjaro, you can start with hill around. The new curriculum has the same content, but the problem is how to implement because my colleagues in the schools do not have the strengths, they do not understand. Our teaching methodology insists on localization of knowledge.

One of them could not speak English and she described in Kiswahili what she had learned from the training:

They taught us how to use resources in English and mathematics, for example how to use stones and sticks in mathematics and to teach how to count. They also taught us how to learn games, game teaching with papers. They also taught us how to use pictures to teach English, and how to use real objects.

According to Cavanagh (2009), the pressure to use English in mathematics and science subjects is a reflection of how much attention those subjects are now receiving in the international sphere, and how nations are struggling to balance their desire to gird students for the global job market against issues of national pride and the desire to preserve and promote the use of a native language.

To summarize, the studies in Nigeria and the observation and interviews in Zanzibar revealed that the use of English in science and mathematics has little to do with issues of quality learning. The evidence thus far indicates that using English in science and mathematics does not contribute to an improvement in learning, and it goes against the solid research findings from many countries around the world that show that learning in a local language improves uptake of knowledge; thus, using English in this context negatively impacts quality education. The learning of science and mathematics has to be regarded as a human right that should be accessible to all, with no barriers in using a language that the child has not mastered. The Seeds of Science/Roots of Reading model aims for deep conceptual understanding in local context and in local languages, and engages students in using literacy in the service of inquiry science with an approach used to measure progress along the way, which brings a human rights perspective to learning science.

10
Conclusions and Recommendations: Local Languages and Knowledge for Sustainable Development

In recent decades, most Sub-Saharan African countries have instituted educational reforms, particularly the development of new curricula; however, implementation has not always been suited to the reality, even if the policies were well thought out. Development aid support from international institutions and international development agencies has not prioritized local languages as languages of instruction. Moreover, contextual and material supports are likely to have poor outcomes when implemented in a foreign language. There is a large body of evidence referenced throughout this book confirming that educational reforms in developing countries have rarely been effectively implemented and have often failed to achieve their objectives. Many mistakes in African implementation get repeated because governments model their changes on previous, faulty changes. Cumulative and comparative research knowledge and experiences could be used to describe the successes and failures of implementation and contribute to improvements in new efforts. As discussed in my previous book (Babaci-Wilhite, 2014), the focus in curriculum changes has been on the reforms and not on the implementation process, therefore there is an urgent need to give the process more attention so that methods can be developed to address them. It is obvious than when the changes are not well planned and structured, strong resistance and unexpected outcomes will be the result (Dyer, 1999). In order to adequately implement a new curriculum, the policy makers and planners must take the school realities into account. A comprehensive approach that considers multi-dimensional factors in education and emphasizes quality of life to solve the problem of poverty and its impact on educational opportunity, outside as well as inside of schools, can serve as a model to resolve the complexities of moving toward robust educational systems; in this effort, attention to the local context is crucial. We need a new generation of visionary, inter-professionally oriented leaders who will place the future of all the children of all the people at the top of their agenda where it belongs. Education would constitute the largest work force in the world, undergirded by a new inter-professional ethic and driven by a common mission – child well-being; such a force could accomplish whatever it set out to do. The potential of such a coalition to influence the various forces that develop policies and programs designed to serve the most vulnerable children is unequaled. This effort would represent a significant change from current policies that do not take the language context into consideration, using English as a LoI even though it is a foreign language in most of the countries in Africa and Asia.

Changes in the classroom cannot succeed if the teachers are not involved in the design of the implementation process. Teacher involvement has been underestimated in Africa and this is part of the reason behind a neglect of the problem of teacher competence in English. My research confirms the point made by many scholars, that lack of qualified teachers is one of the root causes of poor quality learning, especially in rural areas. Other contributors are inadequate planning and absenteeism of both head teachers and teachers, large classes, and a lack of quality support materials.

This main contribution of this book is to demonstrate the links between language choice, quality learning and rights in education. Using a local language satisfies the rights criteria of availability, accessibility, acceptability and adaptability. These should be common to education in all its forms and at all levels. In general terms, quality education corresponds to basic education as set out in the World Declaration on EFA, but must also ensure human rights through localizing education in local language and context. It should take account of the educational, cultural and social background of the students concerned. It demands flexible curricula and varied delivery systems to respond to opportunities of communities and the needs of students in different social and cultural settings.

In many African countries as well as some Asian countries, the goals of rights to education and rights in education are becoming increasingly remote. Development aid has to be integrated with human rights principles to meet the demands of rights in education. With this understanding and with the awareness of the education challenges of nations and millions of people throughout Africa and Asia, the implication is that rights in education remains a distant goal. A Rights-Capability-Based Approach to education becomes imperative in order to overcome obstacles. The motive of aid should not be just to increase economic growth or to accelerate adaptation toward the global North, but rather to spread freedom to the unfree. Therefore the Rights-Capability-Based Approach becomes imperative for rights in education (Babaci-Wilhite & Geo-JaJa, 2011). Local languages need to be valued and preserved, and children need to be prepared for the world in a language that promotes understanding.

Even if the promotion of English is the goal of education for many African and Asian countries, the fact that local LoI would produce better results in the eventual acquisition of English should, logically, inform decisions about language-in-education policy. Therefore, as government officials continue to make decisions based on political popularity rather than seeking optimal learning environments, attempts to make changes at the policy level are unlikely to succeed without the support of the communities themselves (Mazrui, 2003).

Despite the scientifically based evidence, English as a LoI continues to be required at all educational levels in many developing nations and remains the focus of many language aid programs implemented by countries such as the USA and the UK (Samoff, 2009). Therefore, language educators working in developmental contexts ought to question language policies and seek to inform other educators, policy makers and community members of more viable educational alternatives to the current, blind-faith reliance on English. Equally important, the political motives of government officials in these countries must be questioned, when many studies show that the LoI and the local curriculum are prerequisites for equity and quality education.

Perhaps the most important point is that change to a foreign LoI will impair students' learning capacity and therefore reduce the quality of education. A quality education should be regarded as a human right. Fundamental freedom and quality education will not be achieved through the medium of a foreign LoI. This change constitutes a violation of children's rights in education.

Based on the declaration of human rights, children have the right to be educated in a way that contributes to their capacity for individual development. Every African child should have the right to express himself/herself in a language s/he masters best; only then can democracy be achieved. African governments have the responsibility to ensure this language right in education. The analysis and results of this book can form the basis for a renewal of debates in Africa on an educational policy that ensures every child's right to quality education. The use of local languages in Africa is the only way to provide a basis for local development on African terms. This conclusion is based on overwhelming evidence on the importance of using a local language in order to achieve quality learning, reinforce local cultural values and make possible development on local terms.

I have argued that sustainable investment should be made in LoI not only in primary schools but also in secondary school. Also, the performance of students should be monitored when they finish secondary school in order to assess whether they have performed better in school. The costs of such a policy are exaggerated and the benefits for quality learning underestimated. The cost of producing support materials and books has been posed as a major cost barrier, but several studies show that books in local languages can be produced at a reasonable cost. These costs associated with implementing a local language are exaggerated. The cost argument is based on a fundamental fear. An analysis of the costs needs to be explored and tested, taking into account the need for considerable investment in order to improve teachers' language proficiency and the overall quality of education in a non-local LoI. These cost-effectiveness perspectives are very important, since most policy makers argue that cost is the main obstacle to using an African or an Asian language as LoI. However, that assessment does not account for the benefits of self-appreciation, better learning and increased understanding that come with a local LoI. This narrow view of cost effectiveness is inhibiting the implementation of a right in education, which is a prerequisite for equitable global development. What is suggested, therefore, is that sustainable investment should be made in local languages as LoI in primary and post-primary schools and that it should be monitored in order to assess its effectiveness in terms of student achievement. This implementation of a local LoI and curriculum will bring quality "Learning for All" – a right in education in all African and Asian countries.

The spread of English is closely tied to the forces of globalization. The languages of economically dominant countries tend to have a readily acceptable and more powerful place in the world than those of less developed countries. Instead of focusing on improved quality, much attention has been given to reducing the cost of education per student at all levels by increasing class size as advised by the World Bank, whose economists claim that the student/teacher ratio does not affect learning in the interval between 20 and 45 students per teacher. The implication of this, that one teacher may teach 45 students at a time with the same quality as if the number of students was only 20, is highly questionable. The success of implementing new curriculum reform depends on the extent to which policy makers and planners take school realities into account. As shown in the case of Zanzibar, the policy seems to have been driven by political imperatives that had little to do with classroom realities (Babaci-Wilhite, 2013).

The advantages of teaching children in their local languages go beyond academic success to include cultural, emotional, cognitive and socio-psychological benefits – this combines the goals of rights to education and rights in education. The literature has shown that when people feel that they are outsiders or when there exists linguistic alienation of the majority from the education system, social problems and conflicts intensify, thus degrading the intrinsic value contribution of education to cultural identity, nation-building and sovereignty, all of which are essential to development on local terms.

The use of a local language in the educational system imparts self-respect to learners and contributes to the decolonization of local culture. By reinforcing the importance of local languages, one reinforces the interest in local knowledge and skills. As I have argued, this would help students to develop their learning skills and problem solving, which together contribute to the development of critical thinking and observational skills. The acquisition of knowledge and confidence related to use of local languages would contribute to sustainable development. This reconsideration would go in the opposite direction of many of the reforms now taking place in Africa, but it would give a basis for evaluating performance and making a decision about whether African languages improve or negatively impact on learning.

I recommend that a pilot secondary school implement this suggestion as soon as possible in order to base future decisions on scientifically valid results. The results of such a pilot will be important because they will contribute to our understanding of the relationship between LoI and good learning, to rights in education and to children's pride and confidence in their community as well as their ability to understand and engage with the world on their own terms.

Today's reality is that it will take time for governments to put into place a system to develop African languages as LoI. However, this will provide a sustainable benefit for each country. As a result, children of all backgrounds will be able to perform better in school, because children taught in any of the language varieties similar to their MT will have improved learning comprehension compared with those taught in an adopted foreign language such as English. The policy of switching from an African language to a foreign language midway through the schooling process gives the impression that African languages are inferior to foreign languages and that the local language is somehow inadequate for engaging with complex concepts. This reinforces the sense of inferiority of local culture and at the same time is disadvantageous for children of the lowest socio-economic strata who have had little exposure to English at home. The best way to improve education is to take research results into consideration and not let the myth of the market's demand for English as a LoI continue to set the agenda.

Ensuring that aid as well as international partnerships contribute to quality education requires the design of more innovative frameworks that fit the uniqueness and realities of localities. These are some of the roadblocks and challenges confronting donors in their efforts to facilitate the provision of rights, efficiency and efficacy in education (Babaci-Wilhite & Geo-JaJa, 2011). Policy makers are in a position to work toward high-quality education for all as part of a more comprehensive Rights-Based Approach. There is a dire need for new thinking about African and Asian education and its role in development. I am suggesting the need for an educational transformation in Africa and Asia, based on cooperation and sharing of expertise within the continent. This is in line with the 1997 SADC Protocol on Education, which prioritized teacher education and sharing of expertise within the region. This sharing can facilitate cross-national learning and better decisions, acceleration of the pace of changes at lower costs and the most effective use of expertise on the continent. This in turn can bring about educational systems that are better suited to African learning environments and contextualized in Africa's development needs. The dominating discourse associating incorporation of Western knowledge with educational advancement has become embedded in much of African thinking concerning the future of teaching and learning. This needs to be dis-embedded and replaced with local knowledge and local languages.

Another issue for further research is the role of development aid in the education sector and the extent to which it is possible to incorporate a new framework that includes support for local languages and a locally based curriculum within the framework of human rights. This book should encourage both educators and development policy makers to ask why wealthy donor nations such as the USA and the UK spend large amounts of "foreign aid" on the promotion of English in developing countries instead of using it for funding basic literacy acquisition in local dialects and generating quality educational materials in native languages. A possible way forward could be to encourage a wide-scale educational campaign to inform developing communities of how language choice in education can affect personal and economic development. In order to make any change possible, one would need to question both the causes and the effects of such harmful language and educational policies at every level, from government officials and policy makers down to the poorest participants in education.

Bringing technology to the human rights in education agenda

This section will briefly address further recommendations on how to work toward the achievement of two important goals for education in developing countries, the integration of global and local learning in the digital age and the retention of local LoI in the digital age. Further, it will argue that the integration of e-learning into conventional curricula, using local languages and knowledge, should be defined as a global justice tool.

I envision education as a set of processes intended to enhance multicultural dimensions in Glo-local learning where e-learning has to be taken into consideration within education in order to enhance citizenship for the digital 21st century. This is an important departure from most mainstream educational research and practice, because e-learning is used and connects most countries around the world through digital devices. In this conceptualization, reading, writing and language are employed as tools to foster the development of Glo-local knowledge and inquiry in science using the Seeds of Science/Roots of Reading model as described in the previous chapter.

In this discussion, I first examine key assumptions about knowledge that inform mainstream educational research and practice. I then argue for an emphasis on the contextualized dimension of learning as a human right in education. This means accounting for contextualized knowledge and reversing the current trend in the digital age of de-contextualization. This will form a new platform for innovation with a unique mix of local and global knowledge in the deployment of digital technology for learning in a digital age.

Reversing the current trend in the digital age

My first recommendation addresses how to incorporate global and local learning and sustain local languages of instruction in the digital age using the Seeds of Science/Roots of Reading model in developing nations, through its local adaptation and the use of digital devices. The second recommendation is that this sustenance of e-learning in local languages in education based in local knowledge should be defined as a human right in education. My third recommendation explores formal and informal learning and how these facilitate cognitive learning in the development of new knowledge, and will give attention to the conjunction of several aspects to improve the quality of learning through science literacy using e-learning in local languages in education as well as local curriculum.

E-learning grounded in the Seeds of Science/Roots of Reading approach will form a new platform for innovation based on a unique mix of local and global knowledge and will enable content production and distribution in writing, reading, exploring and understanding science as well as the application of multimedia on a scale previously unimaginable (Jewitt, 2008). It will provide teachers and students with the capacity to understand and deploy new technologies for learning by using digital devices, thus providing local learners the opportunity to engage with new media and to participate in vital learning experiences. In Africa and Asia only small minorities have access to digital devices, though Warshchauer and Matuchniak (2010) argue that "the large and growing role of new media in the economy and society serves to highlight their important role in education, and especially in promoting educational equity" (p. 180). Furthermore they argue that "effective deployment and use of technology in schools can help compensate for unequal access to technologies in the home environment and thus help bridge educational and social gaps" (ibid.).

The Seeds of Science/Roots of Reading model would address how reading and writing can be used as tools to support inquiry-based science in developing nations and how to support the implementation of good science teaching in today's complicated curricular landscape through the use of a digital device. To date, the computer has been the digital device used in education to facilitate learning. However, it is revealing to note that 80% of the world's population does not have access to basic infrastructure facilities that are regarded as commonplace in the industrialized world – and that there are reported to be more computers in New York city alone than in the whole of Africa (DFID, 2000 quoted by Crossley & Watson, 2003). "If every city resembled New York, everyone could access the other societys education systems to discover what can be learned that will contribute to improved policy and practice at home" (Arnove, 2003, p. 6). However, the main digital device used in classrooms is no longer a computer.

Lesson learned from using digital learning

In 2012, the US Department of Education and the Federal Communications Commission encouraged the development of digital textbooks and digital learning in K–12 public education. According to the Pew Research Internet Project (PRIP), the rise of e-reading (2012) has increased and in 2013, 50% of US adults own a tablet computer and/or an e-book reader. Furthermore PRIP demonstrated interest in the use of these devices in schools, as most teachers and students emphasize that tablets are:

- much more lightweight than print textbooks
- hold hundreds of textbooks

- reduce environmental impacts by reducing the amount of printing
- increase student interactivity and creativity
- cheaper than print textbooks.

The counter arguments are that tablets are:

- expensive
- too distracting for students
- easy to break
- costly/time-consuming to fix
- quickly outdated as new technologies are released.

Several states in the USA, according to the "Digital Textbook Playbook", have begun transitioning their instructional materials from paper textbooks to digital learning, such as:

- California in 2009 (launched a free digital textbook initiative) with

    1) San Diego United School District distributing 78,000 digital textbooks to teachers and students since 2011 and 26,000 iPads in 2012;
    2) Los Angeles United School District (the second-largest school district in the country) purchasing 655,000 iPads for all K–12 students.

- West Virginia in 2009 (replaced social studies print textbook purchases with digital textbooks).
- Florida in 2015 (mandated that all K–12 instructional materials must be provided in electronic format).
- Georgia state law requiring that electronic copies of K–12 curriculum be made available for use by students.

In the African and Asian context, with no books and no support materials for teaching and learning, tablets can be very useful and adaptable. The American Association of Publishers argues that while the average net unit price of a K–12 print textbook in 2010 was $65, digital textbooks on average cost $50 to $60. The cost of tablets in the USA was on average $489 in 2011 and $343 in 2013, and was projected to be $263 in 2015. The implementation costs for e-textbooks on iPad tablets are higher than new print textbooks due to Wi-Fi infrastructure, the need to train teachers and administrators how to use the technology, and annual publisher fees to continue using e-textbooks.

In developing nations, with high demands for literacy and technology, the use of tablets has a huge potential to improve teaching and learning as long as the content is contextualized. I propose that the Seeds of Science/Roots of Reading model of instruction developed and taught through a tablet would contribute to the sustainability and development of science literacy in Asia and Africa. Its application would be a major step in correcting violations of children's rights in education due to the weak learning environment pointed out in the previous chapters, especially the lack of trained teachers and support materials. This could contribute to a solid curriculum grounded in thorough teacher preparation and quality support materials.

Language is crucial to the learning process inside and outside of schools. E-learning can be a tool of learning in the future as long as the curriculum is grounded in the local context and mediated through a local language. The emphasis should be on indigenous concepts, articulated in their natural environment. I will conclude that education is more than schooling; therefore rethinking in re-teaching is crucial in order to promote social justice locally and globally, and using a digital device will contribute to the achievement of this goal.

My hope is that this book will contribute to a better understanding of the importance of local LoI in learning and to an increasing awareness of the optimal teaching/learning environment that maximizes the performance of children and teachers within the context of their rights in education.

Notes

1 Introduction: The Paradigm Shift in Language Choices in Education

1. I will refer to the Republic of Tanzania as Tanzania (mainland) and Zanzibar to distinguish them; although they are united, Zanzibar has a school system autonomous from that of the Tanzania mainland.
2. "Swahili" itself (in general) refers to culture, people and in other literature it is used synonymously with Kiswahili to mean the "language" (Ismail, 2013).
3. Picture taken by Zehlia Babaci-Wilhite in a school in Dar-es-Salaam, Tanzania.

6 Linguistic Rights for Appropriate Development in Education

1. https://treaties.un.org/Pages/Treaties.aspx?id=4&subid=A&lang=en.
2. Tomasevski, K. (2006). *Human Rights Obligations: Making Education Available, Accessible, Acceptable and Adaptable*, p. 16.
3. Tomasevski, K. (2006). *Human Rights Obligations in Education: The 4-A Scheme*, p. 86.
4. Tomasevski, K. (2001). *Human Rights Obligations: Making Education Available, Accessible, Acceptable and Adaptable*, p. 12.
5. Tomasevski, K. (2001). *Human Rights Obligations: Making Education Available, Accessible, Acceptable and Adaptable*, p. 14.

7 Language-in-Education Policy

1. Tanganyika originally consisted of the British share of the former German colony of German East Africa, which the British took under a League of Nations mandate in 1922 and which was later transformed into a United Nations trust territory after World War II. On April 26th, 1964, Tanganyika joined with the islands of Zanzibar to form the United Republic of Tanganyika and Zanzibar, a new state that changed its name to the United Republic of Tanzania within a year.
2. This class is also known as a "bridge year", taking place after the last year of primary school and before starting Form 1, the first year of secondary school.
3. Picture taken by Maryam J. Ismail in a school in Zanzibar.
4. Programme for Institutional Transformation Research Outreach, a Norwegian-funded program at the University of Dar-es-Salaam.

9   Science Literacy and Mathematics as a Human Right

1. Languages of wider communication, often spoken across national borders.
2. During an interview in February 2012 with the Vice Chancellor at the State University of Zanzibar, I was told that the Zanzibari Kiswahili is the Oxford of Kiswahili (a reference to Oxford English), that is, the purest.
3. Student Performance in National Examinations, a project conducted by the University of Bristol.

Bibliography

Action Aid, Real Aid (2006). *Making Technical Assistance Work*, July 5, 2006, p. 14 (OECD Figures for 2004).

Action Aid, Promoting Rights in Schools (2011). *South Africa Action Aid*, Real Aid 3 Report, 2011, UK.

ADEA (2001). *What works and what's new in education: Africa speaks! Report from a prospective, stocktaking Review of Education in Africa*. Paris: ADEA-International Institute for Educational Planning.

Afflerbach, P., Pearson, P. D. & Paris, S. G. (2008). Clarifying differences between reading skills and reading strategies, *Research Based Classroom Practice*, 61(5), 364–373.

Afolayan, A. (1990). Bilingualism and bilingual education in Nigeria. In: Paulston, C. (Ed.), *International Handbook of Bilingualism and Bilingual Education* (pp. 33–60). New York: Cambridge University Press.

Agnihotri, R. K. (2001). English in Indian Education. In: Daswani, C. J. (Ed.), *Language Education in Multilingual India* (pp. 186–209). New Delhi: UNESCO.

Aikman, S. (1995). Language, literacy and bilingual education. An Amazon people's strategies for cultural maintenance, *International Journal of Educational Development*, 15(4), 411–422.

Akande, A. T. & Salami, L. O. (2010). Use and attitudes towards Nigerian Pidgin English among Nigerian university students. In: Millar, R. & McColl, R. (Eds.), *Marginal Dialects: Scotland, Ireland and Beyond* (pp. 70–89). Aberdeen: Forum for Research on the Languages of Scotland and Ireland.

Akyeampong, E. (2000). Africans in the Diaspora: The Diaspora in Africa, *African Affairs*, (99), 183–215.

Alesina, A. & Dollar, D. (2000). Who gives foreign aid to whom and why? *Journal of Economic Growth*, 5(1), 33–63.

Alesina, A. & Weder, B. (2002). Do corrupt governments receive less foreign aid? *American Economic Review*, 92(4), 1126–1137.

Alidou, H. (2003). Medium of instruction in post-colonial Africa. In: Tollefson, J. W. & Tsui, A. (Eds.), *Medium of Instruction Policies: Which Agenda? Whose Agenda?* (pp. 195–214). Mahwah & London: Lawrence Erlbaum Publishers.

Alidou, H. & Brock-Utne, B. (2011). Teaching practices – teaching in a familiar language. In: Ouane, A. & Glanz, C. (Eds.), *Optimising Learning, Education and Publishing in Africa: The Language Factor. A Review and Analysis of Theory and Practice in Mother-Tongue Bilingual Education in Sub-Saharan Africa* (pp. 159–185). Hamburg/Tunis: UNESCO Institute for Lifelong Learning/Association for the Development of Education in Africa.

Alkire, S. (2002). *Valuing Freedoms: Sen's Capability Approach and Poverty Reduction*. New York: Oxford University Press.

148

Alston, P. & Robinson, M. (2005). Introduction: The challenges of ensuring the mutuality of human rights and development endeavors. In: Alston, P. & Robinson, M. (Eds.), *Human Rights and Development: Towards Mutual Reinforcement* (pp. 1–18). Oxford: Oxford University Press.

Amin, S. (1998). *Spectres of Capitalism: A Critique of Intellectual Fashions.* Translated by Shane Mage. New York: Monthly Review Press.

Amin, S. (2003). *Obsolescent Capitalism: Contemporay Politics and Global Disorder.* London: Zed Books.

Amin, S. (2014). The right to quality education. Translated by Zehlia Babaci-Wilhite. In: Babaci-Wilhite, Z. (Ed.), *Giving Space to African Voices: Rights in Local Languages and Local Curriculum* (pp. 85–93). Rotterdam: Sense Publishers.

Anderson, G. & Aresenault, N. (2002). *Fundamentals of Educational Research.* London: Routledge Farmer of the Taylor & Francis Group.

Andersen, T. B., Harr, T. & Tarp, F. (2006). On US politics and IMF lending, *European Economic Review*, 50(7), 1843–1862.

Appadurai, A. (1990). Disjuncture and difference in the global culture economy, *Theory, Culture & Society*, 7(2), 295–310. Doi: 10.1177/026327690007002017.

Arnove, R. F. (1980). Comparative education and world-systems analysis. The comparative and international education society, *Comparative Education Review*, February. 3 24(1), 48–62. Doi: 10.1086/446090.

Arnove, R. F. (2003). Introduction: Reframing comparative education. The dialectic of the global and the local. In: Arnove, R. F. & Torres, C. (Eds.), *Comparative Education: The Dialectic of the Global and the Local* (pp. 1–22). Lanham, MD: Rowman & Littlefield.

Arnove, R. F. (2009). World-systems analysis and comparative education in the age of Globalization. In: Cowen, R. & Kazamias, A. M. (Eds.), *International Handbook of Comparative Education* (pp. 101–109). New York: Springer Publishing.

Arnove, R. F. (2012). Globalization, diversity, and education. In: Banks, J. A. (Ed.), *Encyclopedia of Diversity in Education* (vol. 2, pp. 1006–1014). Thousand Oaks, CA: Sage.

Arnove, R. & Arnove, A. (1998). *Issues in Post-colonial Languages Policies: Essentialism vs Universalism.* Cape Town: WCCES.

Arnove, R. F., Franz, S., Mollis, M. & Torres, C. A. (2003). Education in Latin America: Dependency, underdevelopment, and inequality. In: Arnove, R. F. & Torres, C. A. (Eds.), *Comparative Education: The Dialectic of the Global and the Local* (pp. 313–337). Lanham, MD: Rowman & Littlefield Publishers.

Asmah, H. O. (1992). *The Linguistic Scenery in Malaysia.* Kuala Lumpur, Malaysia: Dewan Bahasa dan Pustaka, Ministry of Education.

Babaci-Wilhite, Z. (2010). Why is the choice of the language of instruction in which students learn best seldom made in Tanzania? In: Brock-Utne, B., Desai, Z. & Qorro, M. (Eds.), *Educational Challenges in Multilingual Societies: LOITASA Phase Two Research* (pp. 281–305). Cape Town, South Africa: African Minds.

Babaci-Wilhite, Z. (2012a). Local language of instruction for quality learning and social equity in Tanzania. In: Yeung, A. S., Brown, E. L. & Lee, C. (Eds.), *Communication and Language: Barriers to Cross-Cultural Understanding* (vol. 7, pp. 3–23). Charlotte, NC: Information Age Publishing.

Babaci-Wilhite, Z. (2012b). A right based approach to Zanzibar's language-in-education policy. Special Issue on Right Based Approach and Globalization in Education. *World Studies in Education*, 13(2), 17–33.

Babaci-Wilhite, Z. (2013a). An analysis of debates on the use of a global or local language in education: Tanzania and Malaysia. In: Napier, D. B. & Majhanovich, S. (Eds.), *Education, Dominance and Identity* (vol. 3, pp. 121–133). Rotterdam: Sense Publishers.

Babaci-Wilhite, Z. (2013b). The new education curriculum in Zanzibar: The rationale behind it. In: Desai, Z., Qorro, M. S. A. & Brock-Utne, B. (Eds.), *The Role of Language in Teaching and Learning Science and Mathematics: LOITASA Phase Two Research* (pp. 127–151). Cape Town: African Minds.

Babaci-Wilhite, Z. (2014a). *New Perspective on Language & Literacy in Africa and Asia*. A Lead Paper Presented at the 17th Annual National Conference of the Association for Promoting Nigerian Languages and Cultures, July 14th, 2014, Owerri – Nigeria.

Babaci-Wilhite, Z. (Ed.) (2014b). *Giving Space to African Voices: Right in Local Languages and Local Curriculum* (vol. 33, pp. 1–217). Rotterdam: Sense Publishers.

Babaci-Wilhite, Z. (2015). *Local Language of Instruction as a Human Right in Education: Cases from Africa* (vol. 36, pp. 1–154). Rotterdam: Sense Publishers.

Babaci-Wilhite, Z. & Geo-JaJa, M. A. (2011). A Critique and Rethink of Modern Education in Africa's Development in the 21st Century. *Papers in Education and Development (PED), Journal of the School of Education*, (30), 133–154. The United Republic of Tanzania: University of Dar es Salaam.

Babaci-Wilhite, Z. & Geo-JaJa, M. A. (2014). Localization of instruction as a right in education: Tanzania and Nigeria language-in-education's policies. In: Babaci-Wilhite, Z. (Ed.), *Giving Space to African Voices: Right in Local Languages and Local Curriculum* (vol. 33, pp. 3–19). Rotterdam: Sense Publishers.

Babaci-Wilhite, Z., Geo-JaJa, M. A. & Lou, S. (2012a). Education and language: A human right for sustainable development in Africa, *International Review of Education*, 58(5), 619–647.

Babaci-Wilhite, Z., Geo-JaJa, M. A. & Lou, S. (2012b). Human rights in development experience in Africa: The Foreign aid and policy Nexus in OECD and China aid. *World Studies in Education*, 13(1), 79–90.

Babaci-Wilhite, Z., Geo-JaJa, M. A. & Mwajuma, V. (2013). Nordic aid and the education sector in Africa: The case of Tanzania. In: Brown, C. A. (Ed.), *Globalization, International Education Policy, and Local Policy Formation* (vol. 5, pp. 81–106). New York: Springer Publishing.

Bakahwemama, J. (2010). What is the difference in achievement of learners in selected Kiswahili and English-medium primary schools in Tanzania? In:

Brock-Utne, B., Desai, Z. & Qorro, M. A. S. (Eds.), *Educational Challenges in Multilingual Societies: LOITASA Phase Two Research* (pp. 204–228). Cape Town: African Minds.

Baker, D. P. (2009). The invisible hand of world education culture. In: Plank, S. G. & Schneider, B. (Eds.), *Handbook of Education Policy Research*. New York: Routledge.

Baker, D. P. & Le Tendre, G. (2005). *National Differences, Global Similarities: World Culture and the Future of Schooling*. Stanford, CA: Stanford University Press.

Baldwin, D. (1985). *Economic State Craft*. Princeton, NJ: Princeton University Press.

Bamgbose, A. (1976). *Mother Tongue Education: The West African Experience*. London: Hodder and Stoughton Ltd.

Bamgbose, A. (1982). Standard Nigerian English: Issues of identification. In: Kachru, B. (Ed.), *The Other Tongue. English across Cultures* (pp. 99–111). Oxford: Pergamon Press.

Bamgbose, A. (1984). Mother tongue medium and scholastic attainment in Nigeria, *Prospects*, 16(1), 87–93.

Bamgbose, A. (2000). *Language and Exclusion: The Consequences of Language Policies in Africa*. Münster, Hamburg, London: LIT.

Bamgbose, A. (2003). A recurring decimal: English in language policy and planning, *World Englishes*, 22(4), 413–431.

Bamgbose, A. (2005). Mother-tongue education: Lessons from the Yoruba experience. In: Brock-Utne, B. & Hopsen, R. K. (Eds.), *Language of Instruction for African Emancipation: Focus on Post-colonial Contexts and Considerations* (pp. 231–257). Cape Town & Dar es Salaam: CASAS and Mkuki na Nyota Publishers.

Bangladesh Bureau of Educational Information and Statistics (BANBEIS) (1999). *National Education Survey (Post-Primary) 1999*. Dhaka: BANBEIS. Final report.

Barber, J. (2005). *The Seeds of Science/Roots of Reading Inquiry Framework. The Lawrence Hall of Science*. USA: University of California-Berkeley. Accessible at: http://scienceandliteracy.org/sites/scienceandliteracy.org/files/biblio/barber_inquirycycle_pdf_54088.pdf.

Bari, M. (2002). *A Brief History of the Bangla Language Movement*. Accessible at: http://www.cyberbangladesh.org/banglahist.text.

Barrett, A. M., Crossley, M. & Dachi, H. (2011). International collaboration and research capacity building: Learning from the EdQual Experience, *Comparative Education*, 47(1), 25–43.

Batibo, H. (1995). The growth of Kiswahili as language of education and administration in Tanzania. In: Putz, M. (Ed.), *Discrimination through Language in Africa: Perspectives on the Namibian Experience* (pp. 57–80). Berlin: Mouton de Gruyter.

Benson, C. (2010). Language of instruction as the key to educational quality: Implementing mother-tongue-based multilingual education. *Policy Brief*. Stockholm: Swedish International Development Authority.

Benson, C. & Kosonen, K. (Eds.) (2013). *Language Issues in Comparative Education: Inclusive Teaching and Learning in Non-Dominant Languages and Cultures* (vol. 24, pp. 1–314). Rotterdam: Sense Publishers.

Bernstein, B. (1971). On the classification and framing of educational knowledge. In: Young, M. (Ed.), Knowledge and control: New directions in the sociology of knowledge (pp. 47–69). London: Collier-Macmillan.

Berthémlemy, J. C. (2006). Bilateral donors' interest vs. recipients' development motives in aid allocation: Do all donors behave the same? *Review of Development Economics*, 10(2), 179–194.

Bishop, R. & Glynn, T. (1999). *Culture Counts: Changing Power Relations in Education. Whose Knowledge Counts? How Is Knowledge Constructed?* (pp. 225). New Zealand: Dunmore Press.

Boone, P. (1996). Politics and the effectiveness of foreign aid, *European Economic Review*, 40(2), 289–329.

Bostad, I. (2013). Right to education – for all?: The quest for a new humanism in globalization, *World Studies in Education*, 14(1), 7–16.

Bourdieu, P. (1977). *Outline of a Theory of Practice.* Cambridge: Cambridge University Press.

Breidlid, A. (2009). Culture, indigenous knowledge systems and sustainable development: A critical view of education in an African context, *International Journal of Educational Development*, 29, 140–148.

Brenner, M. (2006). Interviewing in educational research. In: Green, J., Camilli, G. & Elmore, P., Skukaukaité A. & Grace, E. (Eds.), *Handbook of Complementary Methods in Education Research* (pp. 357–370). Washington, DC: American Educational Research Association.

Brock-Utne, B. (2000, reprinted in 2006). *Whose Education for All? The Recolonization of the African Mind.* Seoul: Home Publishing Co. Reprinted by Africancbooks.org.

Brock-Utne, B. (2005). The continued battle over Kiswahili as the language of instruction in Tanzania. In: Brock-Utne, B. & Hopson, R. K. (Eds.), *Languages of Instruction for African Emancipation: Focus on Postcolonial Contexts and Considerations* (pp. 51–87). Cape Town: CASAS and Dar es Salaam: Mkupi na Nyota. Also published by African Books Collective in Oxford, UK.

Brock-Utne, B. (2007). Worldbankification of Norwegian Development Assistance to Education. *Comparative Education.* Special Issue on *Global Governance, Social Policy and Multilateral Education*, 43(3), 433–449.

Brock-Utne, B. (2010). Research and policy on the language of instruction issue in Africa, *International Journal of Educational Development*, 30(4), 636–645.

Brock-Utne, B. (2011). Language and inequality: Global challenges to education. BAICE Presidential Address at the 11th UKFIET International Conference, Oxford.

Brock-Utne, B. (2012). Language policy and science: Could some African countries learn from Asian countries? *International Review of Education*, 58(4), 481–503.

Brock-Utne, B. & Garbo, G. (Eds.) (2009). *Language and Power. Implications of Language for Peace and Development.* Dar es Salaam: Mkuki na Nyota. Oxford: ABC, East Lansing: Michigan State University Press.

Brook Napier, D. (2011). Critical issues in language and education planning in twenty first century in South Africa, *US–China Education Review*, 1, 58–76.

Bruthiaux, P. (2002). Hold your courses: Language education, language choice, and economic development, *TESOL Quarterly*, 36, 275–96.

Bryman, A. (2004). *Social Research Methods.* 2nd Edition. Thousand Oaks, CA, USA: Sage Publications, Inc.

Buchert, L. (1994). *Education in the Development of Tanzania 1919–1990.* London, England: Ohio University Press, Eastern African Studies.

Bull, H. & Watson, A. (Eds.) (1984). *European International Society and Its Expansion.* Oxford: Clarendon Press.

Bull, H. & Watson, A. (Eds.) (1995). *The Expansion of International Society.* Oxford: Clarendon Press.

Bunyi, G. (1999). Rethinking the place of African indigenous languages in African education, *International Journal of Educational Development*, 19(4), 337–350.

Carceles, G., Fredriksen, B. & Watt, P. (2001). *Can Sub-Saharan Africa Reach the International Targets for Human Development? Africa Region Human Development* (Working Paper Series No. 9). Washington, DC: The World Bank.

Carney, S., Rappleye, J. & Silova, I. (2012). World culture theory and comparative education, *Comparative Education Review*, 56(3), 366–394.

Carnoy, M. (1999). *Globalization and Education Reform: What Planners Need to Know?* Paris: United Nations Educational, Scientific and Cultural Organization.

Carnoy, M. (2007). *Cuba's Academic Advantage.* Palo Alto, CA: Stanford University Press.

Cashel-Cordo, P. & Craig. S. (1997). Donor preferences and recipient fiscal behavior: A simultaneous analysis of foreign aid, *Economic Inquiry*, 35(3), 653–671.

Cavanagh, S. (2009). Curriculum matters, Malaysia reverts to teaching math and science in native. Accessible at: http://blogs.edweek.org/edweek/curriculum/2009/07/malayonly_math_and_science.html.

Cervetti, G. N., Pearson, P. D., Bravo, M. A. & Barber, J. (2005). *Reading and Writing in the Service of Inquiry-Based Science.* Berkeley: University of California Press.

Cervetti, G. N., Pearson, P. D., Bravo, M. A. & Barber, J. (2006). Reading and writing in the service of inquiry-based science. In: Douglas, R., Klentschy, M. & Worth, W. (Eds.), *Linking Science and Literacy in the K–8 Classroom* (pp. 221–244). Arlington, VA: National Science Teachers Association.

Cervetti, G. N., Barber, J., Dorph, R., Pearson, P. D. & Goldschmidt, P. G. (2012). The impact of an integrated approach to science and literacy in elementary school classrooms, *Journal of Research in Science Teaching*, 49(5), 631–659.

Cervetti, G. N., Pearson, P. D., Barber, J., Hiebert, E. H. & Bravo, M. A. (2007). Integrating Literacy and Science: The Research We Have, the Research We Need. In: Pressley, M., Billman A. K., Perry K., Refitt, K. & Reynolds J. (Eds.), *Shaping Literacy Achievement* (pp. 157–174). NewYork: Guilford.

Chambers, R. (1997). *Whose Reality Counts?: Putting the First Last.* London: Intermediate Technology Development Group Publishing.

Chatterjee, L. (1992). Landmarks in official educational policy: Some facts and figures. In: Rajan, R. S. (Ed.), *The Lie of the Land* (pp. 300–308). New Delhi: Oxford.

Chisholm, L. & Leyendecker, R. (2008). Curriculum reform in post-1990s sub-Saharan Africa, *International Journal of Educational Development*, 28(2), 195–205.

Chisholm, L. & Steiner-Khamsi, G. (Eds.) (2009). *South–South Co-operation in Education and Development.* New York: Teachers College Press.

Chowdhury, A. (2001). *Hope Not Complacency: The State of Primary Education in Bangladesh.* Dhaka: Universal Publication Limited.

Colclough, C. (1993). Primary schooling in developing countries: The unfinished business. In: Allsop, T. & Brock, C. (Eds.), *Oxford Studies in Comparative Education* (vol. 3, pp. 47–57). Wallingford, CT: Triangle Books.

Coleman, H. (2009a). Teaching other subjects through English in three Asian nations: A review. Unpublished report for British Council East Asia.

Coleman, H. (2009b). Are "International Standard Schools" really a response to globalisation? Paper presented at the International Seminar "Responding to Global Education Challenges", held at Universitas Negeri Yogyakarta, Indonesia, May 19th, 2009.

Coleman, H. (2009c). Teaching other subjects through English in two Asian nations: Teachers' responses and implications for learners. In: Powell-Davies, P. (Ed.) *Access English EBE Symposium: A Collection of Papers* (pp. 63–87). Kuala Lumpur: British Council East Asia.

Collier, P. (2008). Poverty and Development: An IMF seminar, quoted by Olav Lundstøl at Norad Annual conference in Oslo, October 15th, 2010.

Cooke, J. & Williams, E. (2002). Pathways and labyrinths: Language and education in development, *TESOL Quarterly*, 36, 297–322.

Crossley, M. (2000). Bridging cultures and traditions in the reconceptualization of comparative and international education, *Comparative Education*, 36(3), 319–332.

Crossley, M. (2010). Context matters in educational research and international development, partnerships and capacity building in small states experiences, *Prospects, Quarterly Review of Comparative Education*, 40(4), 421–429.

Crossley, M. & Watson, K. (2003). *Comparative and International Research to Education: Globalisation, Context and Difference.* London and New York: Routledge.

Dalgard, C.-J. & Hansen, H. (2000). On aid, growth, and good policies, *Journal of Development Studies*, 37(6), 17–42.

Danish International Development Agency (DANIDA) (1992). *Evaluation Report: The School Maintenance Project Tanzania.* Denmark: Nordic Consulting Group A/S of Greve Denmark.

David, M. K. (2007). Changing language policies in Malyasia: Ramifications and Implications. Paper presented at the 2nd International Conference on Language, Education and Diverstiy held at the Unversity of Waikato-Hamilton, New Zealand, November 21st–24th, 2007.

De Feiter, L. P., Vonk, H. & Van der Akker, J. (1995). *Towards More Effective Teacher Development in Southern Africa.* Amsterdam: VU University Press.

Desai, Z. (2003). A case for mother tongue education? In: Brock-Utne, B., Desai, Z. & Qorro, M. A. (Eds.), *Language of instruction in Tanzania and South Africa (LOITASA)* (pp. 45–68). Dar Es Salaam: E & D Limited.

Dewey, J. (1938). *Experience and Education.* New York: Collier Books (Collier edition first published 1963).

Dikshit, O. (1974): Students should be encouraged to think in a foreign language, *Journal of the Nigerian English Studies Association,* 6(2), 47–52.

Dreze, J. & Sen, A. (2002). *India: Development and Participation.* Oxford: Oxford University Press.

Dyer, C. (1999). Researching the implementation of educational policy: A backward mapping approach, *Comparative Education,* 35(1), 45–61.

Easterly, W. (2008). Can the west save Africa? *Journal of Economic Literature,* 47(2), 373–447.

Education for All (EFA) Assessment. (2000). Global synthesis. UNESCO Publishing. Oxford University Press. Accessible at: http://www.unesco.org/new/en/education/themes/leading-the-international-agenda/education-for-all/.

Education for All. (2005). EFA Global Monitoring Report. UNESCO Publishing. Oxford University Press. Accessible at: http://www.unesco.org/new/en/education/themes/leading-the-international-agenda/efareport/reports/2005-quality/.

EFA (2009). Global Monitoring Report. UNESCO Publishing. Oxford University Press.

Egbe, G. B., Bassey, B. U. & Otu, R. O. (2004). Beyond academic reading: Encouraging popular reading for life-long learning, *Literacy and Reading,* 8(1), 54–58.

Emenanjo, E. N. (2008). *Languages and the National Policy on Education: Implications and Prospects.* Accessible at: http://fafunwafoundation.tripod.com/fafunwafoundation/id9.html.

Erstad, O., Gilje, O., Sefton-Green, J. & Vasbo, K. (2009). Exploring "learning lives": Community, identity, literacy and meaning, *Literacy,* 43(2), 100–106.

European Network on Debt and Development (EURODAD) (2008). *Turning the Tables: Aid and Accountability Under the Paris Framework: A Civil Society Report.* Brussels, Belgium: Eurodad Publisher.

Fafunwa, B. A. (1990). Using national languages in education: A challenge to African educators. In: UNESCO/UNICEF, *African Thoughts on the Prospects of Education for All.* Selections from papers commissioned for the regional consultation on Education for All. Dakar, November 27th–30th, pp. 97–110.

Fafunwa, B. A., Macauley, J. I, Sokoya, J. A. (Eds.) (1989). *Education in Mother Tongue: The Ife_Primary Education Research Project.* Ibadan: University Press Limited.

Ferraro, V. (1996). Dependency theory: An introduction. In: Secondi, G. (Ed.), *The Development Economics Reader* (pp. 58–64). London: Routledge.

Foucault, M. (1988). Power, moral values, and the intellectual. An interview with Michel Foucault by Michael Bess, *History of the Present*, (4), 1–2. Accessible at: http://variazionifoucaultiane.blogspot.no/2011/12/power-moral-values-and-intellectual.html.

Frank, A. G. (1966). The development of underdevelopment – from Volume 18, 1966, *Monthly Review* reprint.

Fredriksen, B. (2005). Keeping the promise: What is holding up achieving primary education for all African children? *Finance & Development*, Washington, DC: International Monetary Fund.

Freire, P. (1970, Reprinted in 1993). *Pedagogy of the Oppressed.* London: Penguin Books, Ltd.

Freire, P. (1977). Guinea Bissau: Record of an ongoing experience, *Convergence* (Toronto, International Council for Adult Education), X (4), 11–28.

Fullan, M. (1991). *The New Meaning of Educational Change* (2nd Ed.). London: Cassell.

Galabawa, J. (2002). The language strategy of education and development. Paper presented to the launch workshop of Language Instruction in Tanzania and South Africa (LOITASA), Morogoro, Tanzania.

Galabawa, J. & Lwaitama, A. F. (2005). A comparative analysis of performance in Kiswahili and English as languages of instruction at secondary level in selected Tanzania schools. In: Brock-Utne, B., Desai, Z. & Qorro, M. A. (Eds.), *LOITASA Research in Progress* (pp. 139–159). Dar es Salaam: KAP Associates.

Galtung, J. (1971). A structural theory of imperialism, *Journal of Peace Research*, 8(2), 81–117.

Gandhi, M. K. (1954). *Medium of Instruction* (B. Kumarappa, Trans.). Ahmedabad: Navjivan Publishing House.

Geo-JaJa, M. A. (2004). Decentralization and privatization of education in Africa: Which option for Nigeria? *Special Issue of International Review of Education*, 50(3–4), 309–326.

Geo-JaJa, M. A. (2006). Educational decentralization, public spending, and social justice in Nigeria, *International Review of Education*, 52(1–2), 129–153.

Geo-JaJa, M. A. (2009). *Can Globalization in Nigeria's Niger Delta Be Humanized for Integration and Development? Language and Power. Implications of Language for Peace and Development.* Dar es Salaam: Mkuki na Nyota; Oxford: ABC; East Lansing: Michigan State University Press.

Geo-JaJa, M. A. (2013). Education localization for optimizing globalization's opportunities. In: Majhanovich, S. & Geo-JaJa, M. A. (Eds.), *Economics, Aid and Education: Implications for Development* (pp.159–182). Rotterdam: Sense Publishers.

Geo-JaJa, M. A. & Azaiki, S. (2010). Development and education challenges in the Niger Delta in Nigeria. In: Geo-JaJa, M. A. & Majhanovich, S. (Eds.),

Education, Language, and Economics: Growing National and Global Dilemmas (pp. 53–69). Rotterdam: Sense Publishers.

Geo-JaJa, M. A. & Mangum, G. (2000). Donor aid and human capital formation: A neglected priority in development. In: Koehn, P. & Ojo, O. J. B. (Eds.), *Making Aid Work: Strategies and Priorities for Africa at the Turn of the Twenty-First Century* (pp. 97–115). Lanham, MD: University Press of America.

Geo-JaJa, M. A. & Mangum, G. (2003). Economic adjustment, education and human resource development in Africa: The case of Nigeria, *International Review of Education*, 49(3–4), 293–318.

Geo-JaJa, M. A. & Yang, X. (2003). Rethinking globalization in Africa. *Chimera*, 1(1), 19–28.

Geo-JaJa, M. A. & Zajda, J. (2005). Rethinking globalization and the future of education in Africa. In: Zajda, J. (Ed.), *International Handbook of Globalization, Education and Policy Research*. Amsterdam: Kluwer Academic Publishers.

Geo-JaJa, M. A. (2006). Educational decentralization, public spending, and social justice in Nigeria, *International Review of Education*, 52(1–2), 129–153.

Gibbons, A. (1985). *Information, Ideology and Communication: The New Nations' Perspectives on an Intellectual Revolution*. Lanham, MD: University Press of America.

Gill, S. K. (2005). Language Policy in Malaysia: Reversing Direction, *Language policy*, 4(3), 241–260.

Goffman, E. (1959). *The Presentation of Self in Everyday Life*. New York: Doubleday and Anchor Books.

Goldenberg, C. (2008). *Teaching English Language Learners, What the Research Does – and Does Not – Say*. American Educator. Accessible at: http://www.aft.org/sites/default/files/periodicals/goldenberg.pdf.

Goulet, D. (1988). Development . . . or liberation. In Wilbur, C. K. (Ed.), *1988. The Political Economy of Development and Underdevelopment* (pp. 379–386). New York: Random House.

Habwe, J. (2009). The Role of Kiswahili in the Integration of East Africa, *The Journal of Pan African Studies*, 2(8). Accessible at: http://www.jpanafrican.com/docs/vol2no8/2.8_RoleOfKiswahiliInTheIntegration.pdf.

Haddad, W. (1995). *Education Policy-Planning Process: An Applied Framework; Fundamentals of Educational Planning*. No. 51. Paris: UNESCO/IIEP.

Haki Elimu (2009). *HakiElimu Lugha*, Video on Science classroom on Global warming issues; uploaded on February 27th, 2009. Accessible at: http://www.youtube.com/watch?v=3uxaBP8gsG4&feature=related.

Hamm, B. (2001). A human rights approach to development, *Human Rights Quarterly*, 23(4), 1005. Doi: 10.1353/hrq.2001.0055.

Hassan, A. (1988). *Bahasa Sastera Buku Cetusan Fikiran*. Kuala Lumpur: Dewan Bahasa dan Pustaka.

Hassan, R. (2003). Globalization, literacy and ideology, *World Englishes*, 22(4), 433–448.

Hayman, R. (2005). Are the MDGs enough? Donor perspectives and recipient visions of education and poverty reduction in Rwanda. Presented at the UKFIET Oxford Conference on Education and Development, September.

Heneveld, W. & Craig, H. (1996). Schools count: World Bank project designs and the quality of primary education in Sub-Saharan Africa. World Bank Technical Paper, no. 303, The World Bank, Washington, DC.

Heng, C. S. & Tan, H. (2006). English for mathematics and science: Current Malaysian language-in-education policies and practices, *Language and Education*, 20(4), 306–321.

Heugh, K. (2004). Is multilingual education really more expensive? In: Pfaffe, J. F. (Ed.), *Making Multilingual Education a Reality for All: Operationalizing Good Intentions* (pp. 212–233). (Proceedings of 3rd International Conference of DALEST and 1st Malawian Language Symposium, Mangochi, Malawi, 30 August–03 September, 2003), Centre for Language Studies, Zomba.

Heugh, K. (2006). Cost implications of the provision of mother tongue and strong bilingual models of education in Africa. In: Ouane, A. & Glanz, C. (Eds.), *Optimising Learning and Education in Africa – the Language Factor: A Stock-Taking Research on Mother Tongue and Bilingual Education in Sub-Saharan Africa* (pp. 255–289). Hamburg: UNESCO Institute for Lifelong Learning (UIL) and Association for the Development of Education in Africa (ADEA).

Heugh, K. (2011). Theory and practice – language education models in Africa: Research, design, decision-making and outcomes. In: Ouane, A. & Glanz, C. (Eds.), *Optimising Learning, Education and Publishing in Africa: The Language Factor. A Review and Analysis of Theory and Practice in Mother Tongue Bilingual Education in Sub-Saharan Africa* (pp. 105–157). Hamburg/Tunis: UNESCO Institute for Lifelong Learning/Association for the Development of Education in Africa.

Higgins, E. (2004). Marching to different drums: Exploring the gap between policy and practice in the education sector in Uganda. Unpublished PhD Thesis. University of Bristol, UK.

Hornberger, N. & Vaish, V. (2009). Multilingual language policy and school linguistic practice: Globalization and English language teaching in India, Singapore and South Africa. *Compare: A Journal of Comparative and Interventional Education*, 39(3), 305–320.

Imam, S. R. (2007). *English as a Global Language and the Question of Nation-Building Education in Bangladesh*. Doi: 10.1080/03050060500317588, pp. 471–486.

Ishumi, A. (1985). Educational research in Tanzania – some issues. In: Ishumi, A. G., Komba, D., Bwatwa, Y. M., Mosha Herme, J., Katunzi, N. & Mahenge, B. S. T. (Eds.), *Educational Research in Tanzania* (pp. 2–23) (Papers in Education and Development; 10). University of Dar es Salaam. Department of Education.

Ishumi, A. G. M. (1992). Colonial forces and ethnic resistance in African education. In: Schleicher, K. & Kozma, T. (Eds.), *Ethnocentrism in Education* (pp. 121–138). Germany: Frankfurt am Main: Peter Lang.

Islam, M. (2011). A pragmatic language policy in relation to English: Bangladesh contexts. In: M. S. Thirumalai, M. S. (Ed.) *Language in India: Strength for Today and Bright Hope for Tomorrow* (vol. 11, pp. 47–57). ISSN 1930–2940.

Ismail, M. J. (2007). English in Zanzibar: Triumphs, trials and tribulations: NAWA, *Journal of Language and Communication*, 1(2), 1–13.

Ismail, M. J. (2013). *Being and Becoming a Teacher of English: A Study of Teacher Educators and Their Students in Postcolonial Zanzibar*, PhD thesis. Monash University, Australia.

Ismail, M. J. (2014). Infusing a rights-based approach in initial teacher education in postcolonial Zanzibar: Critical insiders' perspectives. In: Babaci-Wilhite, Z. (Ed.), *Giving Space to African Voices: Rights in Local Languages and Local Curriculum* (vol. 33. pp. 173–195). Rotterdam: Sense Publishers.

Jewitt, C. (2008) *Technology, Literacy and Learning: A Multimodal Approach*, London: Routledge.

John, J. C. (2010). What is the difference in the quality of education provided by government and private primary schools in Tanzania? A comparative study. In: Brock-Utne, B., Desai, Z. & Qorro, M. A. S. (Eds.), *Educational Challenges in Multilingual Societies: LOITASA Phase Two Research* (pp. 229–253). Cape Town: African Minds.

Kachru, B. (1986). The power and politics of English, *World Englishes*, 5(2–3), 121–140.

Kachru, B. (1990). World Englishes and applied linguistics, *World Englishes*, 9(1), 3–30.

Kachru, B. (1997). English as an Asian language, *Links & Letters*, 5, 89–108.

Kadeghe, M. (2003). In defense of continued use of English as the language of instruction in secondary and tertiary education in Tanzania. In: Brock-Utne, B., Desai, Z. & Qorro, M. A. S. (Eds.), *Language of Instruction in Tanzania and South Africa* (LOITASA) (pp. 150–170). Cape Town: African Minds.

Kamwangamalu. N. M. (2003). Social change and language shift in South Africa. *Annual Review of Applied Linguistics*, 23, 225–242.

Killick, T. (2004). Politics, evidence and the new aid agenda. *Development Policy Review*, 22(1), 29–53.

Killick, T. (2008). Taking control: Aid management policies in least developed countries (Background paper for the United Nations Conference on Trade and Development UNCTAD). *The Least Developed Countries Report: Growth, Poverty and Terms of Development Partnership 2008*. Accessible at: www.unctad.org/sections/ldc_dir/docs/ldcr2008_Killick_en.pdf.

Kimizi, M. M. (2012). Language and self-confidence: Examining the relationship between language of instruction and students' self-confidence in Tanzanian secondary education. PhD dissertation, University of Oslo. Institute for Educational Research, Faculty of Education.

Ki-Zerbo, J. (1990). *Educate or Perish: Africa Impass and Prospects;* BREDA with WCARO. Dakar, Senegal: UNESCO-UNICEF (Western-Africa).

Klees, S. (2010). Aid, development and education, *Current Issues in Comparative Education*, 13(1), 3–28.

Klees, S. & Qargha, O. (2013). The economics of aid: Implications for education and development. In: Majhanovich, S. & Geo-JaJa, M. A. (Eds.), *Economics, Aid and Education: Implications for Development* (pp. 15–29). Rotterdam: Sense Publishers.

Knack, S. (2001). Aid dependence and the quality of governance: A cross-country empirical analysis. World Bank Working Paper. Washington, DC.

Kosonen, K. (2008). Language-in-education policies in Southeast Asia. Keynote paper presented during SEAMEO Workshop on "Using Mother Tongue as Bridge Language of Instruction in Southeast Asian Countries: Policy, Strategies and Advocacy", Bangkok, February 19th–20th, 2008.

Kosonen, K. (2010). Cost-effectiveness of first language-based bilingual & multilingual education. In: Pushker, K. (Ed.), *Multilingual Education in Nepal Kathmandu* (pp. 27–33). Centre-Nepal: Language Development.

Kuczynski, P. & Williamson, J. (2003). *After the Washington Consensus: Restarting Growth and Reform in Latin America.* Washington, DC: Institute for International Economics.

Lancaster, C. (2007). The Chinese aid system. Center for Global Development. Accessible at: http://www.cgdev.org/files/13953_file_Chinese_aid.pdf.

Loona, S. (1996). Monolinguismo es curable (Monolingualism can be cured)? In: De Heer-Dehue, J. (Ed.), *Intercultural Education and Education of Migrant Children: Practice and Perspectives* (pp. 129–154). Comenius Action 2 Seminar, Oegstgeest, the Netherlands, October 25th–26th, 1996, organized for the European Platform for Dutch Education. Utrecht. Stichting Promotie Talen.

Lwaitama, A. F. (2004). Local knowledge and indigenous science: Nyerere's philosophy of education and poverty alleviation. In: Galabawa, J. & Norman, A. (Eds.), *Education, Poverty and Inequality* (pp. 169–197). Dar es Salaam, BERRIPA Project, Faculty of Education.

Lwaitama, A. F. & Rubagumya, C. M. (1990). The place of English in Tanzanian education system: The political and economic dimensions to the current and alternative language policy options, *Journal of Linguistics and Language Education*, 4(4), 143.

Machingaidze, T., Pfukani, P. & Shumba, S. (1998). The quality of education: Some policy suggestions based on a survey of schools, *SACMEQ Policy Research* No. 3: Zimbabwe. Paris: IIEP.

Majhanovich, S. (2013a). English as a tool of neo-colonialism and globalization in Asian contexts. In: Hébert, Y. & Abdi, A. (Eds.), *Critical Perspectives in International Education* (pp. 249–262). Rotterdam: Sense Publishers.

Majhanovich, S. (2013b). How the English language contributes to sustaining the neo-liberal agenda: Another take on the strange non-demise of neo-liberalism. In: Majhanovich, S. & Geo-JaJa, M. A. (Eds.), *Economics, Aid and Education: Implications for Development* (pp. 79–96). Rotterdam: Sense Publishers.

Majhanovich, S. (2014). Neo-liberalism, globalization, language policy and practice issues in the Asia-Pacific Region, *Asia-Pacific Journal of Education*, 34(2), 168–463.

Malekela, G. (2003). English as a medium of instruction in post-primary education in Tanzania: Is it a fair policy to the learner? In: Brock-Utne, B., Desai, Z. & Qorro, M. A. S. (Eds.), *LOITASA Language of Instruction in Tanzania and South Africa* (pp. 102–112). Dar es Salaam: E & D Limited Publishers.

Makalela, L. (2005). We speak eleven tongues. Reconstructing multilingualism in South Africa. In: Brock-Utne, B. & Hopson, R. K. (Eds.), *Languages of*

Instruction for African Emancipation: Focus on Postcolonial Contexts and Considerations (pp. 147–175). Cape Town: CASAS and Dar es Salaam: Mkuki na Nyota.

Marjorie, L. (2010). Language policy and oppression in South Africa. CSQ Issue: 6.1 (Spring 1982) Poisons and Peripheral People: Industrial Hazards.

Mason, M. (2011). What underlies the shift to a modality of partnership in educational development co-operation? *International Review of Education.* 57(3), 443–455. Doi: 10/1007s11159-011-9219-7.

Mazrui, A. (1997). The World Bank, the language question and the future of African education, race & class, *A Journal for Black and Third World Liberation,* 38(3), 35–49.

Mazrui, A. (2002). The English language in African education: Dependency and decolonization. In: Tollefson, J. (Ed.), *Language Policies in Education: Critical Issues* (pp. 267–281). Mahwah, NJ: Lawrence Erlbaum Associates.

Mazrui, A. (2003). *English in Africa: After the Cold War.* Sydney: Multilingual Matters Ltd.

Mazui, A. (2003a). Towards re-Africanizing African Universities: Who killed intellectualism in the post-colonial era? *Turkish Journal of International Relations,* 2(3&4), 135–163

Mbaabu, I. (1996). *Language Policy in East Africa.* Nairobi: Educational Research and Publications (ERAP).

McGinn, N. F. (1997). Not decentralization but integration. In: Watson, K., Modgil, C. & Modgil, S. (Eds.), *Power and Responsibility in Education.* London: Cassell.

McKinley, R. & Little, R. (1979). The U.S. aid relationship: A test of the recipient need and the donor interest models, *Political Studies,* 27(2), 236–250.

Meyer, J. W. (1971). Economic and political effects on national educational enrollment patterns, *Comparative Education Review,* 15(1), 28–43.

Meyer, J. W. (1992). *School Knowledge for the Masses: World Models and National Primary Curricular Categories in the Twentieth Century.* London: Falmer.

Ministry of Education (1974). *Qudrat-e-Khuda Education Commission 1974.* Dhaka: Ministry of Education.

Ministry of Education (1988). *Mofiz Uddin Education Commission 1988.* Dhaka: Ministry of Education.

Ministry of Education (1997). *Shamsul Haque Education Committee 1997.* Dhaka: Ministry of Education.

Ministry of Education (1998). *National Education Policy 1998.* Dhaka: Ministry of Education.

MoEC (Ministry of Education and Culture) (2002). Education projects and programs. Dar-es-Salaam Office.

MoECS (Ministry of Education, Culture and Sports) Zanzibar (2003) Mid-term Review of the Zanzibar Education Master Plan – 1996–2001 – Interim Report. Presented to the Ministry of Education, Culture and Sports Government of Zanzibar and UNESCO. Dar-es-Salaam Office.

MoECS (Ministry of Education, Culture and Sports) (2005). Zanzibar Education Policy. Stown Town: Working Group on Development of Zanzibar Education Policy (February).

MoEVT (Ministry of Education and Vocational Training) (2007). *Zanzibar Education Development Consortium* (ZEDCO).

MoEVT (Ministry of Education and Vocational Training) (2011). Zanzibar, ten years of education development 2000–2010. Prepared by the Ministry of Education and Vocational Training, Zanzibar.

Ministry of Foreign Affairs (2003). *Human Rights in Swedish Foreign Policy.* Government Communication 2003/4: 20, 118, Sweden.

Ministry of Foreign Affairs (2004). Fighting poverty together. A comprehensive development policy. Report No. 3 (2003–2004) to the Storting. Oslo, Norway.

Ministry of Foreign Affairs (2007). *A World for All.* Priorities of the Danish Government for Danish Development Assistance 2008–2012. Copenhagen, Denmark.

Mori, I. & Baker, D. P. (2010). The origin of universal shadow education: What the supplemental education phenomenon tells us about the postmodern institution of education, *Asia Pacific Education Review*, 11(1), 36–48.

Mosha, H. J. (2006). *Planning educational system for excellence.* Dar es Salaam: E & D Limited.

Mulokozi, M. M. (1991). English versus Kiswahili in Tanzania's secondary education. In: Blommaert, J. (Ed.), *Swahili Studies: Essays in Honour of Marcel van Spaandonck* (pp. 7–16). Gent, Belgium: Academie Press.

Mulokozi, M. M. (2000). *Kiswahili as a National and International Language.* Tanzania: Institute of Kiswahili Research, University of Dar-es-Salaam.

Mulokozi, M. M., Rubanza, Y., Senkoro, F. E. M. K., Kahigi, K. K. & Omari, S. (2008). *Report of the 3rd Secondary School Kiswahili Language Textbooks Writing Workshop,* Morogoro: [Arts Books Workshop], held at Amabilis Catholic Center, November 14th–17th, 2008. TUKI/TATAKI.

Mwinsheikhe, H. M. (2002). Science and the language barrier: Using Kiswahili as a medium of instruction in Tanzania secondary schools as a strategy of improving student participation and performance in science (Report 10/1 Education in Africa). Oslo: Institute for Educational Research.

Mwinsheikhe, H. M. (2003). Overcoming the language barrier: An in-depth study of the Tanzania secondary school science teachers' initiatives in coping with the English–Kiswahili dilemma in the teaching–learning process: A research in the making. Paper presented to the NETREED conference, December 8th–10th, 2003, Gausdal, Norway.

Mwinsheikhe, H. M. (2009). Spare no means: Battling with the English/Kiswahili dilemma in Tanzanian secondary school classrooms. In: Brock-Utne, B. & Skattum, I. (Eds.), *Languages and Education in Africa – A Comparative and Transdisciplinary Analysis* (pp. 223–237). Oxford: Symposium Books.

Nassor, S. & Mohammed, K. A. (1998). The quality of education: Some policy suggestions based on survey of schools, Southern Africa Consortium For Monitoring Educational Quality (SACMEQ) Policy Report 4, Paris: Ministry of Education Zanzibar and UNESCO.ED 480283. Accessible at: htpp://unesdoc.unesco.org/images/0011/001151/1151373eo.pdf.

National Bureau of Statistics: Tanzania Official Website (2004). Accessible at: http://www.nbs.go.tz/.

National Policy on Education (2004). *Federal Republic of Nigeria.* NPE, 4th Edition. Lagos: Federal Government Press.

Nelson, P. & Dorsey, E. (2003). At the nexus of human rights and development: New methods and strategies of global NGOs, *World Development,* 31(12), 2013–2026. Doi: 10.1016/j.worlddev.2003.06.009.

Ngugi wa Thiong'o, J. (1993). *Moving the Centre: The Struggle for Cultural Freedoms.* Islington, London: James Curry Ltd.

Ngugi wa Thiong'o, J. (1994). *Decolonising the Mind: The Politics of Language in African Literature.* London and Portsmouth: James Curry Ltd.

Nkamba, M. & Kanyika, J. (1998). The quality of education: Some policy suggestions based on a survey of schools. SACMEQ Policy Research No. 5: Zambia. Paris: IIEP.

Norad (Norwegian Agency for Development) (2010). Publications on Norwegian development aid. Accessible at: http://www.norad.no/en/tools-and-publications/publications.

Norad (Norwegian Agency for Development) (2012). Tanzania: For several decades Tanzania has been one of the most important countries for Norwegian development co-operation. Available at: http://www.Norad.no/en/countries/africa/tanzania.

Nussbaum, M. (1998). Capabilities and human rights, *Fordham Law Review,* 66(2), 273–300.

Nussbaum, M. (2000). *Women and Human Development: The Capabilities Approach.* Cambridge: Cambridge University Press.

Nussbaum, M. (2011). *Creating Capabilities: The Human Development Approach.* Harvard: Harvard University Press.

Nyerere, J. K. (1967, Reprinted in 1968). Education for self-reliance. In Ujamaa, (Ed.), *Essays on Socialism* (pp. 44–76). Dar-es-Salaam: Oxford University Press.

Obanya, P. (1999). Popular fallacies on the use of African languages in education. Social dynamics, Special Issue: *Language and Development in Africa,* 25(1), 81–100.

Obuasi, I. (2014). English language in anuku dialect environment: A query on language of instruction, *Journal of International Society of Comparative Education, Science and Technology of Nigeria,* 14(1), 78–87.

Ocampo, J. (2004). Latin America's growth and equity frustration during structural reforms, *Journal of Economic Perspective,* 18(2), 67–88.

Odora, C. H. (Ed.) (2002). Stories of the hunt – who is writing them? In: *Indigenous Knowledge and the Integration of Knowledge Systems. Towards a Philosophy of Articulation* (pp. 237–257). Claremont, South Africa: New Africa Education.

OECD (Organisation for Economic Co-operation and Development) (2005). The Paris declaration on aid effectiveness and the Accra agenda for action. Paris. Accessible at: http://www.oecd.org/dataoecd/11/41/34428351.pdf.

OECD (Organisation for Economic Co-operation and Development) (2006). *Integrating Human Rights into Development: Donor Approaches, Experiences and Challenges.* Paris: OECD the Development Dimension Series.

OECD (Organisation for Economic Co-operation and Development) (2008). Human rights and aid effectiveness: Key actions to improve inter-linkages. Paris. Accessible at: http://www.oecd.org/dataoecd/13/63/43495904.pdf.

OECD (Organisation for Economic Co-operation and Development) (2010). Donors' mixed aid performance for 2010 sparks concern, Development Assistance Committee. Accessible at: http://www.oecd.org/document/20/0, 3343,en_2649_34447_44617556_1_1_1_37413,00.html.

OECD (Organisation for Economic Co-operation and Development) (2011). Common ground between South–South and North–South Co-operation principles. Accessible at: http://www.oecd.org/dac/dacglobalrelations/ Common%20ground%20between%20South-South%20and%20North-South%20co-operation.pdf.

OECD (Organisation for Economic Co-operation and Development) (2011). *Paris Declaration of Aid Effectiveness 2005–10: Progress in Implementing the Paris Declaration.* Paris: OECD Publishing.

Office of the High Commissioner for Human Rights (1996–2014). What are human rights? Online OHCHR Website. Accessible at: http://www.ohchr. org/EN/Pages/WelcomePage.aspx.

Ogunsuji, Y. (2003). A critical assessment of educational and political language policy documents in Nigeria. In: Oyeleye, L. & Olateju, M. (Eds.), *Reading in Language and Literature* (pp. 153–168). Nigeria: Obafemi Awolowo Univerisity Press Limited.

Okanlawon, K. (1987). *Emergent Issues in Nigerian Education,* vol. 1. Lagos: Joja Press Limited.

Okonkwo, C. E. (1983). Language development in Nigerian education – 1800 to 1980, *Journal of Research in Curriculum,* 1(2), 78–92.

Okonkwo, J. I. (2014). Appropriate language in education: The strategy for national development in Nigeria. In: Babaci-Wilhite, Z. (Ed.), *Giving Space to African Voices: Rights in Local Languages and Local Curriculum* (vol. 33. pp. 131–146). Rotterdam: Sense Publishers.

Olarenwaju, A. O. (2008). *Using Nigerian Languages as Media of Instruction to Enhance Scientific and Technological Development: An Action Delayed.* Accessible at: http://fafunwafoundation.tripod.com/fafunwafoundation/ id10.html.

Omojuwa, J. (2005). Laying a strong foundation for higher level of reading achievement: Problems and prospects, *Journal of Applied Literacy and Reading,* 2, 7–15.

Osborne, E. (2002) Rethinking Foreign Aid, *Cato Journal,* 22(2), 297–316.

O'Sullivan, M. (2002). Reform implementation and the realities within which teachers work: A Namibian case study, *Compare,* 32(2), 219–237.

Osungbemiro, N. R., Olaniyan, R. F., Sanni, R. O. & Olajuyigbe, A. O. (2013). Use of Indigenous Languages of Media of Instruction in Teaching of Biology (A case study of some selected secondary schools in Ondo West

Local Government, Ondo State, Nigeria). Cambridge Business & Economics conference, July 2–3, 2013. ISBN: 9780974211428.

Othman, H. (2008). Democracy and language in Tanzania. In: Brock-Utne, B. & Garbo, G. (Eds.), *Language and Power* (pp. 1–9). Dar-es-Salaam: Mkuki na Nyota, Oxford: African Books Collective and East Lansing: MI: State University of Michigan Press.

Ouane, A. (1990). National languages and mother tongues, *UNESCO Courier*, 43(7), 27–29.

Ouane, A. & Glanz, C. (Eds.) (2006). *Optimising Learning and Education in Africa – the Language Factor: A Stock-Taking Research on Mother Tongue and Bilingual Education in Sub-Saharan Africa.* Hamburg: UNESCO Institute for Lifelong Learning (UIL) and Association for the Development of Education in Africa (ADEA).

Panitchpakdi, S. (2006). Secretary-General of UNCTAD, on United Nations Day for South–South Co-operation. New dynamics of South–South development.

Pearson, P. D. (2007). An endangered species act for literacy education, *Journal of Literacy Research*, 39(2), 145–162.

Pearson, P. D. & Hiebert, E. H. (2013). Understanding the common core state standards. In: Morrow, L., Shanahan, T. & Wixson, K. K. (Eds.), *Teaching with the Common Core Standards for English Language Arts: What Educators Need to Know* (Book 1: pp. 1–21). New York: Guilford Press.

Pearson, P. D., Moje, E. & Greenleaf, C. (2010). Literacy and science: Each in the service of the other, *Science*, 328, 459–463.

Pennycook, A. (1994). *The Cultural Politics of English as an International Language.* Harlow, Essex, UK: Longman Group Limited.

Pennycook, A. (1995). English in the world/the world in English. In: Tollefson, J. (Ed.) *Power and Inequality in Language Education* (pp. 34–58). Cambridge: Cambridge University Press.

Pennycook, A. (2000). Language, ideology, and hindsight. In: Ricento, T. (Ed.), *Ideology, Politics, and Language Policies: Focus on English* (pp. 49–65). Philadelphia: John Benjamins.

Phillipson, R. (1992). *Linguistic Imperialism.* Oxford: Oxford University Press.

Phillipson, R. (Ed.) (2000). *Rights to Language,* Lawrence Erlbaum Associates. ISBN 0-8058-3835-X.

Phillipson, R. (2002). Global English and local language policies. In: Kirkpatrick, A. (Ed.), *Englishes in Asia: Communication, Identity, Power and Education* (pp. 7–28). Melbourne: Language Australia.

PISA (Programme for International Student Assessment) (2006). *Science Competencies for Tomorrow's World.* Paris: Organisation for Economic Co-operation & Development (OECD).

Pogge, T. (2002). *World Poverty and Human Rights.* Cambridge: Polity Press.

Postman, N. (1979). *Teaching as a Conserving Activity.* New York: Delacorte.

Prah, K. K. (2003). Going native: Language of instruction for education, development and African emancipation. In: Brock-Utne, B., Desai, Z. & Qorro, M. A. S (Eds.), *Language of Instruction in Tanzania and South Africa* (LOITASA) (pp. 14–34). Dar es Salaam: E & D Publishing.

Prah, K. K. (2005). Language of instruction for education, development and African emancipation. In: Brock-Utne, B. & Kofi Hopson, R. (Eds.), *Languages of Instruction for African Emancipation: Focus on Postcolonial Contexts and Considerations* (pp. 23–87). Cape Town: CASAS and Dar es Salaam: Mkuki na Nyota.

Prah, K. K. & Brock-Utne, B. (Eds.) (2009). *Multilingualism – An African Advantage. A Paradigm Shift in African Languages of Instruction Policies.* Cape Town, South Africa: CASAS.

Press Reference (2009). Sw-Ur-Tanzania Press, Media, TV, Radio, Newspapers – television, circulation, stations, papers, number, print, freedom. Accessible at: http://www.pressreference.com/Sw-Ur/Tanzania.html.

Prophet, R. & Dow, J. (1994). Mother tongue language and concept development in science. A Botswana case study, *Language, Culture and Curriculum*, 7(3), 205–217.

Psacharopoulos, G. (1989). Why educational reforms fail: A comparative analysis, *International Review of Education*, 35(2), 179–195.

Qorro, M. A. S. (2003). Going native: Language of instruction for education, development and African emancipation. In: Brock-Utne, B., Desai, Z. & Qorro, M. A. S. (Eds.), *Language of Instruction in Tanzania and South Africa (LOITASA).* Dar es Salaam: E&D Limited.

Qorro, M. A. S. (2004). Popularising Kiswahili as the language of instruction through the media in Tanzania. In: Brock-Utne, B., Desai, Z. & Qorro, M. A. S. (Eds.), *Researching the Language of Instruction in Tanzania and South Africa (LOITASA)* (pp. 93–116). Zanzibar: African Minds.

Qorro, M. A. S. (2005). Parents' views on the medium of instruction in Tanzania. In: Brock-Utne, B., Desai, Z. & Qorro, M. A. S. (Eds.), *LOITASA Research in Progress* (pp. 96–124). Dar es Salaam: KAD Associates.

Qorro, M. A. S. (2009). Parents' and policymakers' insistence on foreign languages as media of education in Africa: Restricting access to quality education – for whose benefit? In: Brock-Utne, B. & Skattum, I. (Eds.), *Languages and Education in Africa – a Comparative and Transdisciplinary Analysis* (pp. 57–83). Oxford: Symposium Books.

Rabin, A. (2011). Language of instruction in Tanzanian schools: Creating class divides and decreasing educational standards. Think Africa Press. Accessible at: http://thinkafricapress.com/tanzania/language-instruction-tanzanian-schools-creating-class-divides-and-decreasing-educational-s.

Rafiu, K. A. (2012). Halting the tide of language loss in Niger State, Nigeria. In: Kuupole, D. D., Bariki, I. & Yennah, R. (Eds.), *Cross-Currents in Language, Literature and Translation.* Benin: Universitie Bilingue.

Ramanathan, V. (2004). *The English–Indigenous Divide: Postcolonial Language Politics and Practice.* Clevedon, England: Multilingual Matters.

Ramanathan, V. (2005). Ambiguities about English: Ideologies and critical practice in indigenous-medium college classrooms in Gujarat, India, *Journal of Language, Identity and Education*, 4(1), 45–65.

Robeyns, I. (2005). The capability approach: A theoretical survey, *Journal of Human Development*, 6(1), 93–114.

Robeyns, I. (2006). Three models of education rights: Rights, capabilities and human capital, *Theory and Research in Education*, 4(1), 69–84.

Rodrik, D. (2006). Goodbye Washington consensus, Hello Washington confusion? A review of the World Bank's economic growth in the 1990s: Learning from a decade of reform, *Journal of Economic Literature*, 44(4), 973–987.

Rogan, J. M. (2007). An uncertain harvest: A case study of implementation of innovation, *Journal of Curriculum Studies*, 39(1), 97–121.

Rogan, J. M. & Grayson, D. J. (2003). Towards a theory of curriculum implementation with particular reference to science education in developing countries, *International Journal of Science Education*, 25(10), 1171–1204.

Roy-Campbell, Z. M. & Qorro, M. A. S. (Eds.) (1997). *Language Crisis in Tanzania. The Myth of English versus Education*. Dar es Salaam: Mkuki naNyota Publishers.

Rubagumya, C. (1991). Language promotion for educational purposes: The example of Tanzania, *International Review of Education*, 37(1), 67–68.

Rubagumya, C. (2000). Language as a determinant of quality. In: Galabawa, J., Senkoro, F. E. M. K. & Lwaitama, A. F. (Eds.), *The Quality of Education in Tanzania: Issues and Experiences* (pp. 121–134). The United Republic of Tanzania: University of Dar es Salaam, Faculty of Education.

Rubagumya, C. (2003). English medium primary schools in Tanzania: A new "linguistic market" in education? In: Brock-Utne, B., Desai, Z. & Qorro, M. A. S. (Eds.), *Language of Instruction in Tanzania and South Africa (LOITASA)* (pp. 149–169). Dar-es-Salaam: E&D Limited.

Rwanda (2005). Census 2002. 3ème recensement général de la population et de lhabitat du Rwanda au 15 août 2002. Caracteristiques socio-culturelles de la population. Analyse des resultats. Kigali: Ministe're des Finances et de la Planification Economique/Commission Nationale de Recensement/ Service National de Recensement.

Rwantabagu, H. (2011). Tradition, globalisation and language dilemma in education: African options for the 21st century, *International Review of Education*, 57 (3–4), 457–475.

Ryen, A. (2004). Ethical issues. In: Seale, C., Giampietro, G., Gubrium J. F., Silverman, D. (Eds.), *Qualitative Research Practice* (pp. 217–229). London: Sage.

Said, E. (1991). *Orientalism*. Harmondsworth: Penguin.

Said, O. M. (2003). Orientation secondary courses and the language of instruction in Zanzibar educational system: An evaluation study, Master thesis, University of Oslo. Institute for Educational Research, Faculty of Education.

Samoff, J. (1999). No teacher guide, no textbooks, no chairs: Contending with crisis in African education. In: Arnove, R. F. & Torres, C. A. (Eds.), *Comparative Education: The Dialect of the Global and the Local* (pp. 393–431). Lanham, MD: Rowman & Littlefield.

Samoff, J. (2003, re-edited 2007). Institutionalizing international influence. In: Arnove, R. F. & Torres, C. A. (Eds.), *Comparative Education: The Dialectic of the Global and the Local* (pp. 52–91). Lanham, MD: Rowman & Littlefield Publishers.

Samoff, J. (2009). Foreign aid to education: Managing global transfers and exchanges. In: Chisholm, L. & Steiner-Khamsi, G. (Eds.), *South–South Cooperation in Education and Development* (pp. 123–156). New York: Teachers College Press.

Samuelson, B. L. & Freedman S. W. (2009). Language policy, multilingual education, and power in Rwanda. *Language Policy (2010)*, 98, 191–215.

Schultz, T. (1963). *The Economic Value of Education.* New York: Columbia University.

Schweisfurth, M. (2006). Global and cross-national influences on education in post-genocide Rwanda. *Oxford Review of Education*, 32(5), 697–709.

Scoppio, G. (2002). Common trends of standardisation, accountability, devolution and choice in the educational policies of England, U.K., California, U.S.A, and Ontario, Canada, *Comparative Education*, 2(2), 130–141.

SEAMEO (2008). Workshop Report: SEAMEO – World Bank Project on Using Mother Tongue as Bridge Language of Instruction in Southeast Asian Countries: Policy, Strategies and Advocacy. Bangkok: Southeast Asian Ministers of Education Organisation.

Selbervik, H. (2006a). Donor dilemmas and economic conditionality in the 1990s: When Tanzania is as good as it gets. Paper presented at XIV International Economic History Congress, Helsinki, Session 86.

Selbervik, H. (2006b). Nordic exeptionalism in development assistance? Aid policies and the major donors: The Nordic Countries, CMI Report 2006: 8. Oslo: CHR.

Semali, L. M. (2009). Indigenous pedagogies and language for peace and development. In: Brock-Utne, B. & Garbo, G. (Eds.), *Language and Power. The Implications of Language for Peace and Development* (pp. 196–207). Dar-es-Salaam: Mkupi na Nyota.

Semali, L. M. & Mehta, K. (2012). Science education in Tanzania: Challenges and policy responses, *International Journal of Educational Research*, 53(2012), 225–239.

Sen, A. (1999). *Development as Freedom.* New York: Knopf.

Sen, A. (2004). Elements of a theory of human rights, *Philosophy and Public Affairs*, 32(4), 315–356.

Senghor, L. S. (1976). *Authenticité et Négritude.* Kinshasa. Unpublished lecture.

Senkoro, F. E. M. K. (2004). Research and approaches to the medium of instruction in Tanzania: Perspectives. In: Brock-Utne, B., Desai, Z. & Qorro, M. A. S. (Eds.), *Researching the Language of Instruction in Tanzania and South Africa* (pp. 42–56). Cape Town: African Minds.

Shaheed, F. (2005). The politics of English and Bangla in Bangladesh. *The Daily Star*, March 23rd. Accessible at: http://www.thedailystar.net/2005/03/25/d50325061366.

Shor, I. & Freire, P. (1987). *A Pedagogy for Liberation: Dialogues on Transforming Education.* Westport: CT. Bergin & Garvey Publishers, Inc. An imprint of Greenwood Publishing Group, Inc.

Silova, I. (2010). Private tutoring in Eastern Europe and Central Asia: Policy choices and implications, *Compare: A Journal of Comparative and International Education*, 40(3), 327–344.

Silverman, D. (2011). *Interpreting Qualitative Data* (4th ed.). Thousand Oaks, CA: Sage.

Skutnabb-Kangas, T. (2000). Linguistic human rights and teachers of English: In: Hall, J. K & Eggington, W. G (Eds.), *The Socio Politics of English Language Teaching* (pp. 22–44). Clevedon: Multilingual Matters Ltd.

Skutnabb-Kangas, T. & Heugh, K. (Eds.) (2012). *Multilingual Education and Sustainable Diversity Work: From Periphery to Center.* London: Routledge.

Skutnabb-Kangas, T. & Phillipson, R. (1995). *Linguistic Human Rights: Overcoming Linguistic Discrimination.* New York: Walter de Gruyther.

Sleeter, C. (2001). *Culture, Difference, and Power (CD-ROM edition).* New York: Teachers College Press.

Southern African Development Community (1997). SADC Protocol on Education (1997). Review of the status and capacities for the implementation of the protocol on education and training.

Soyoye, F. A. (2010). Language in Nigerian society and education: Globalization perspective. In: Kuupole, D. D. & Bariki, I. (Eds.), *Applied Social Dimensions of Language Use and Teaching in West Africa* (pp. 28–34). Ghana: The University Press.

Spreen, C. & Vally, S. (2006). Education rights, education policies and inequality in South Africa, *International Journal of Educational Development,* 26(4), 352–362.

Srinivasa, R. (2008). India's Language Debates and Education of Linguistic Minorities. Economic and Political Weekly. September 6, 2008.

Steflja, I. (2012). The Costs and Consequences of Rwanda's Shift in Language Policy. *AfricaPortal,* Backgrounder No. 30, 31 May 2012. Accessible at: http://www.africaportal.org/articles/2012/05/31/costs-and-consequences-rwanda%E2%80%99s-shift-language-policy.

Stiglitz, J. (2002). *Globalization and Its Discontent.* New York and London: Norton Press.

Svensson, J. (2003). Why conditional aid does not work and what can be done about it? *Journal of Development Economics,* 70(2), 381–402.

Tandon, Y. (2008). *Ending Aid Dependence.* Geneva: Fahuma Books, South Center.

Tomasevski, K. (2003). *Education Denied: Costs and Remedies.* London: Zed Books.

Tomasevski, K. (2006). *Human Rights Obligation in Education: The 4-A Scheme.* Nijmegen: Wolf Legal Publishers.

UN [United Nations] (1948). *Preamble: The Universal Declaration on Human Rights.* Adopted by the General Assembly on December 10th. Geneva: United Nations.

UN [United Nations] (1992). General assembly declaration of the rights of persons belonging to national or ethnic, religious and linguistic minorities of 1992.

UN [United Nations] (2003). The human rights-based approach to development cooperation: Towards a common understanding among the UN agencies. Accessible at: http://www.undg.org/?P=221.

170   *Bibliography*

UN [United Nations General Assembly] (2011). Interim report of the Special Rapporteur on the right to education to the 66th session of the UN General Assembly, 2011. New York: UN General Assembly document A/66/269.

UNESCO [United Nations Educational, Scientific and Cultural Organization] (2005). *Children Out of School: Measuring Exclusion from Primary Education.* Montreal: UNESCO, Institute for Statistics.

UNESCO [United Nations Educational, Scientific and Cultural Organization] (2005). *Convention on the Protection and Promotion of the Diversity of Cultural Expressions.* Paris: UNESCO.

UNESCO [United Nations Educational, Scientific and Cultural Organization] (2007). Access to Secondary School. Accessible at: http://unesdoc.unesco.org/images/0019/001900/190007e.pdf.

UNESCO (United Nations Educational, Scientific and Cultural Organization) (2009). Overcoming inequality: why governance matters? Accessible at: http://www.unesco.org/new/en/education/themes/leading-the-international-agenda/efareport/reports/2009-governance/.

UNESCO [United Nations Educational, Scientific and Cultural Organization] (2010). *EFA Global Monitoring Report 2010. Education for All: Reaching the Marginalized.* Paris: Oxford University Press.

UNESCO [United Nations Educational, Scientific and Cultural Organization] (2010). Asia-Pacific secondary education system review, Series No. 2, Françoise Caillods, Access to Secondary Education.

UNDP [United Nations Development Programme] (2006). *Human Development Report.* New York: United Nations.

UNICEF [United Nations International Children's Emergency Fund] (2003). *The State of the World's Children 2004: Girls, Education and Development.* New York: UNICEF.

UNRISD [United Nations Research Institute for Social Development] (2000). Visible hands: Taking responsibility for social development. Geneva: Switzerland.

URT [United Republic of Tanzania] (2004). Tanzania demographic and health survey. Accessible at: http://dhsprogram.com/pubs/pdf/FR173/FR173-TZ04-05.pdf.

URT [United Republic of Tanzania] (2009). *Economic Survey 2008.* Ministry of Planning and Economic Affairs. The United Republic of Tanzania: Dar es Salaam.

Unterhalter, E. (2003). Education, capabilities and social justice. Background paper for EFA Global Monitoring Report 2003/2004, Paris: UNESCO.

Uvin, P. (2004). *Human Rights and Development.* Bloomsfield: Kumarian Press.

Vaidik, V. (1973). *Remove English: Why and How?* New Delhi: The All India Remove English Committee.

Vavrus, F. (2002). Postcoloniality and English: Exploring language policy and the politics of development in Tanzania, *TESOL Quarterly,* 36(3), 373–397.

Vavrus, F. & Bartlett, L. (2012). Comparative pedagogies and epistemological diversity: Social and materials contexts of teaching in Tanzania, *Comparative Education Review,* 56(4), 634–658.

Vedpratap V. (2004). Ambiguities About English: Ideologies and Critical Practice in Vernacular-Medium College Classrooms in Gujarat, India. *Journal of
Language, Identity and Education*, 4(1), 45–65.

Verspoor, A. & Wu, K. B. (1990). Textbooks and educational development.
Educational and employment division, population and human resources
department, PHREE Background Paper Series No. PHREE/90/31, The World
Bank, Washington, DC.

Viswanathan, G. (1989), *Masks of Conquest: Literary Study and British Rule in
India*. New York: Columbia University Press.

Vuzo, M. (2007). Revisiting the language policy in Tanzania: A comparative
study of geography classes taught in Kiswahili and English. Doctoral dissertation. University of Oslo, Institute for Educational Research, Faculty of
Education.

Vuzo, M. (2009). Revisiting the language policy in Tanzania: A comparative
study of geography classes taught in Kiswahili and English. In: Brock-Utne,
B., Qorro, M. A. S. & Desai, Z. (Eds.), *LOITASA: Reflecting on Phase I and
Entering Phase II* (pp. 102–123). Dar es Salaam, Tanzania: E & D Publications.

Wallerstein, I. (1974). *The Modern World-System*. New York: Academic Press.

Wallerstein, I. (1980). *The Modern World-System, vol. II: Mercantilism and
the Consolidation of the European World-Economy, 1600–1750*. New York:
Academic Press.

Warschauer, M., & Matuchniak, T. (2010). New technology and digital worlds:
Analyzing evidence of equity in access, use, and outcomes. *Review of
Research in Education*, 34(1), 179–225.

Waun, X. & Geo-JaJa, M. A. (2013). Internationalization of African Higher
Education: Characteristics and Determinants. *World Studies Education Journal*, Special Issue.

Webley, K. (2006). Mother tongue first: Children's right to learn in their own
languages. *Id21 insights education* vol. 5, September. Accessible at: http://
www.id21.org/insights/insights-ed05/index.html.

Whitelaw, K. 2007). *Rwanda Reborn*, U.S. News and World Report 142, no. 14.
April 23.

Wolff, H. (2006). *Optimizing Learning and Education in Africa – the Language
Factor. A Stock-Taking Research on Mother Tongue and Bilingual Education
in Sub-Saharan Africa*. Association for the Development of Education in
Africa (ADEA): UNESCO Institute for Education – Deutsche Gesellschaft
für Technische Zusammenarbeit. ADEA 2006 Biennial Meeting (Libreville,
Gabon, March 27th–31st, 2006), pp. 26–55.

Wong, L. (1995). Constructing a public popular education in Sao Paulo, Brazil,
Comparative Education Review, 39(1), 120.

World Bank (2000). *Bangladesh: Education Sector Review*. Washington, DC:
World Bank.

World Bank (2003). Bangladesh data profile. Accessible at: http://devdata.
worldbank.org/external/.

World Bank (2008). The evolving allocative efficiency of education aid:
A reflection on changes in aid priorities to enhance aid effectiveness.

Paper prepared for the World Bank High-level Group Meeting on EFA, December 16th–18th, 2008, Oslo, Norway. Washington, DC: World Bank.

Wortham, S. & Jackson, K. (2012). Relational education: Applying Gergen's work to educational research and practice. *Psychological Studies,* 57(2), 164–171.

Young, M. (2008). *Bringing Knowledge Back in: From Social Constructivism to Social Realism in the Sociology of Education.* London: Routledge.

Zajda, J. & Geo-JaJa, M. A. (Eds.) (2010). *The Politics of Education Reforms: Globalization, Comparative Education and Policy Research.* Dordrecht: Springer Publishers.

Zalkapli, A. Z. (2010). Government scraps teaching of maths and science in English, *Malaysian Insider.* Accessible at: http://www.themalaysianinsider.com/index.php/malaysia/31709-government-scraps-teaching-of-maths-and-science-in-english.

Zanzibar Basic Education of Improvement Project (ZABEIP) (2007). A World Bank Group project (approved for April 24th, 2007 – July 31st, 2013). Accessible at: http://www.worldbank.org/projects/P102262/zanzibar-basic-education-improvement-project?lang=en.

Zanzibar Education Development Plan (ZEDP) (1996–2006). Accessible at: ZEMP: http://www.dgmarket.com/tenders/np-notice.do~3990092.

Zanzibar Education Development Programme (2007). *Education Situation Analysis June 2007.* Ministry of Education and Vocational Training (MoEVT). Zanzibar Education Development Consortium (ZEDCO).

Zanzibar Education Development Programme (2007). *Proposed Programs and Activities for the Zanzibar Education Development Plan (2008–2015).* Accessible at: http://planipolis.iiep.unesco.org/upload/Tanzania%20UR/Zanzibar/Zanzibar-Planning-for-ZEDP.pdf

Zanzibar Education Development Programme (ZEDP) (2008/09–2015/16). Accessible at: http://planipolis.iiep.unesco.org/upload/Tanzania.

Zanzibar Education Master Plan (ZEMAP) (2006). Situation analysis of children in Tanzania. Accessible at: http://books.google.no/books?id=JEQUAQAAIAAJ&q=ZEMAP+2006&dq=ZEMAP+2006&hl=no&sa=X&ei=PO7aT_jNLseE4gTTyYHNCg&ved=0CDcQ6AEwAA.

Zanzibar Education Sector Review (2003). Zanzibar education policy. Accessible at: http://planipolis.iiep.unesco.org/upload/Tanzania%20UR/Zanzibar/Zanzibar%20Education%20Policy%20Final.pdf.

Zanzibar Strategy for Growth and Reduction of Poverty (ZSGRP) (2007). Revolutionary Government of Zanzibar, p. 2. Accessible at: http://books.google.com/books?id=uJe2AAAAIAAJ.

Index

Note: Locators followed by 'n' refer to notes.